Ginkgo

Elixir of Youth

MODERN MEDICINE
FROM AN ANCIENT TREE

by Christopher Hobbs

Recycle
Conserve

This book is printed on Simpson 60lb recycled paper .

BOTANICA PRESS

Look for other books available from
Botanica Press

by **Christopher Hobbs**

In natural food stores throughout the U.S.

The Herbs and Health Series:

Echinacea! The Immune Herb
Usnea: The Herbal Antibiotic
Medicinal Mushrooms
Natural Liver Therapy
Vitex: The Women's Herb
Milk Thistle: The Liver Herb
Valerian, The Relaxing and Sleep Herb
Foundations of Health
Kombucha, Manchurian Tea Mushroom
Handbook for Herbal Healing

———————————

6th Printing – December, 1995
Copyright February, 1991
by Christopher Hobbs

Michael Miovic, editor
Beth Baugh, copy editor
Mark Johnson, illustrator
Christopher Hobbs, photographs
Cover Photo © 1995 by Steven Foster

Botanica Press
10226 Empire Grade
Santa Cruz, CA 95060

TABLE OF CONTENTS

INTRODUCTION

Since the dawn of time, people have dreamed of an Elixir of Immortality. Of course, nothing can make us live forever, but nature has given us a remarkable herbal remedy that can slow down the aging process and help make us more healthy and energetic in our later years. This natural medicine comes from the ginkgo tree, and both ancient healers and modern scientists have found it to be effective. Curious? This book explains how ginkgo can—

- Improve memory and brain function
- Protect the heart and restore blood circulation
- Heal hearing and vision problems
- Fight common allergic reactions
- Help preserve general health and vitality

For many people the word *ginkgo* brings to mind an odd-looking tree that often grows in cities and can smell something like rancid butter. But what few people realize is that the mild-mannered (if not mild-odored!) ginkgo is a biological Super Tree: the species has survived almost unchanged for over 150 million years, and some ginkgos live to be over 1,000 years old. In fact, ginkgos are so hardy that a solitary ginkgo was the only tree to survive the atomic blast in Hiroshima. You can still see this tree alive today, standing near the epicenter of the blast, a mute testament to the ginkgo's remarkable ability to survive.[1]

The ginkgo also has amazing medicinal properties. The Chinese, who have always recognized the tree's unique stamina and longevity, have used ginkgo nuts for thousands of years as a remedy for cancer, venereal disease, asthma, lung weakness and congestion, impaired hearing and to increase sexual energy and generally promote longevity. Today the ancient ginkgo has sparked renewed interest throughout the world because medical researchers have isolated chemical compounds from ginkgo that show startling effects in humans. These compounds regulate blood flow to the brain, legs and other extremities and control levels of various neuro-transmitters in the brain, thus helping to counteract memory loss, depression and lack of alertness which may develop in old age. These same compounds also block a substance called *platelet activating factor* (PAF) which, by over-stimulating the immune system, may lead to conditions such as asthma, toxic shock from bacterial poisons and perhaps even atherosclerosis and stroke.

This book will summarize all of these modern, scientific findings on ginkgo and explain them in understandable terms. Drawing on my extensive experience and research as an herbalist, I will also give you practical information and specific instructions on the following:

• What ailments and conditions ginkgo is good for

• How much ginkgo to take and for how long

• What the best ginkgo products are and how to use them

• How ginkgo can be used as a healthy and nutritious dietary supplement

For the more technically-minded, I have included a special appendix full of detailed, scientific information on the botany, chemistry, pharmacology and cultivation of ginkgo. However, note that the information on cultivation may be of interest not only to scientists, but also to those of you who are interested in growing, processing or preparing your own ginkgo medicine. Some herbalists feel that a personal connection with an herb increases its ability to heal, and that besides the purely biochemical action of a plant, the interaction between plants and humans can evolve into a healing session on other, perhaps energetic levels as well. Scientists and doctors who embrace the allopathic model of medicine disagree with this belief, because for them only the objectively measurable effects of an herbal medicine are valid. However, there is no need for dispute here; I will respect both points of view and leave the reader to make his or her own choice. Whether you believe in the traditional wisdom concerning ginkgo, or the modern research—or both—in any case you should find a wealth of thought-provoking and convincing information in this book.

Figure 1

LIST OF AILMENTS THIS BOOK COVERS
(And Where to Find Information About Them and Ginkgo)

Since the first question many people want answered is "Can ginkgo help me?," I am placing the list of conditions and ailments covered in this book right up front. If you or someone you know is suffering from any of these conditions, maybe ginkgo can help. Read on or jump straight to the section indicated to find out if ginkgo is right for you. Also be sure to read the section on Medicinal Preparations (pages 54 -57), where I discuss dosage information and the different kinds of ginkgo products that are available. It is important to use ginkgo—like any dietary supplement or medicine—correctly.

THE HISTORY OF GINKGO

The ginkgo, or maidenhair tree, is one of the oldest living species on this planet. Ginkgo has flourished almost unchanged since 150 million years ago (during the Mesozoic period, when dinosaurs roamed the Earth), and its ancestors can be traced back to 250 million years ago. That is why some scientists call ginkgo "the living fossil." Botanists who study the ancient evolution of plants through their fossil remains have discovered that during prehistoric times ginkgo trees lived in many parts of the world. One of the largest ancient ginkgo forests was on the banks of what is now the Columbia River, near Vantage, Washington. The petrified remains of these forests can still be seen today. During the last ice age, however, ginkgos nearly became extinct; they survived only in China and other parts of Asia—where they stayed until at least

"Ginkgo has flourished almost unchanged since 150 million years ago."

1,000 years ago, when ginkgo trees were planted around monasteries in Japan. (These trees are still living.)

Today there are thought to be no truly wild ginkgo trees left— or perhaps, as some botanists say, only in the forests of eastern China. Ironically, though, in these times of heavy deforestation, ginkgos are now being preserved by the human hand: they are planted as hardy shade trees in many places in the world, especially in cities, partly because they are so resistant to insects, bacteria, viruses, pollution and even old age.

The name *ginkgo* may come from the Chinese *Sankyo* or *Yin-*

kuo (Yin Guo), meaning "hill apricot" or "silver fruit."[2] But although the orange fruits of ginkgo do look something like apricots, these fruits are vastly different from apricots. The small fruits, as many people know, can smell like rancid butter. Kaempfer, the German traveler and surgeon who was the first Westerner to write about ginkgo (in 1712), used the Japanese name *Ginkyo.*. As for the original Chinese name, it is thought that either Kaempfer misspelled it, or it was incorrectly transcribed from his writings by later translators.[3,4]

The Latin name *Ginkgo biloba* L. (formerly *Salisburia adiantifolia* Sm.) was bestowed in 1771 by Linnaeus, the famous Swedish botanist. The word *biloba* means "two lobes" and describes the young leaves of ginkgo which are distinctly two-lobed [see Figure 1]. As an ornamental, ginkgo was introduced into England in 1754 and into America around 1784.

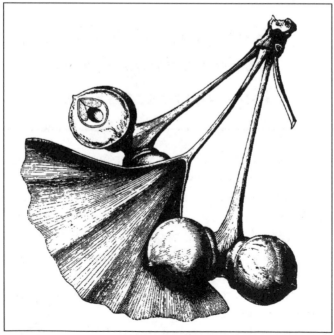

Figure 2

Ginkgo

TRADITIONAL USES OF GINKGO

Although modern medical research focuses mainly on the leaves of ginkgo, the ginkgo fruits and nuts have been used in China since time immemorial as a delicacy and tonic food. Ancient Chinese texts record ginkgo's use as a medicinal agent as far back as five thousand years ago.[5]

Ginkgo fruits, known as *Pak-Ko* in Chinese markets or herb shops, are harvested in the fall, especially after the first frost. The nuts are then cooked and used in porridge and other dishes and are considered a delicacy during weddings and feasts.[6] A traditional method of preparing the nuts for eating is to place the whole fruits, including the orange outer pulp, in a tub of water, allowing them to decompose and ferment. Although the smell exuded during fermentation is highly objectionable (to say the least!), this process is probably essential to producing a safe food and medicine.[7] For although ginkgo nuts that have been roasted, boiled as a tea or cooked in some other way have a very long history of safe use, note well that the *raw* nuts are reported to be toxic, especially to children (one fatality has been reported).[8a] A B6 antivitamin (4'-methoxypyridoxine) may be the cause,[8b] but is probably deactivated by cooking. For this antivitamin to significantly interfere with B6 absorption or utilization, the raw nuts must be eaten over a period of time.

Ginkgo nuts (or seeds, as they are also called) often taste like giant pine nuts. I recently sampled a pot of ginkgo nut porridge in the Emperor Restaurant (the first herbal restaurant on the West Coast, which is now unfortunately defunct), and the waitress was so kind as to give me the following recipe for ginkgo nut porridge, which has been in her family for hundreds of years.

Ginkgo Nut Porridge

Take one cup of rice and 10 to 15 ginkgo nuts, cook in 2.5 cups of water over slow heat, until tender. Remove ginkgo nuts, blend rice until creamy, then add ginkgo nuts. Warm and serve. Add honey, butter or olive oil to taste.

The Japanese are fond of eating ginkgo nuts after meals as a digestive aid. I have eaten ginkgo nuts after hearty meals of rice and vegetables, and I've always found the experience to be beneficial as well as tasty. The nutritional value of ginkgo nuts is good, and they contain very little fat—which makes them an excellent food for the weight-conscious. This is especially true because the fat in ginkgo seeds is mostly unsaturated or monosaturated fatty acids,[9] which are thought to be the most beneficial kinds of oils. Table 1 gives the nutritional profile of ginkgo nuts.

TABLE 1

FOOD VALUE AND DIGESTIBILITY
OF GINKGO NUTS

Ginkgo Nut Food Profile [10,11]

Starch	62-68%
Sucrose	6%
Protein	11-13%
Fat	1.5-3.0%
Fiber	1.0%
Ash	3.4%

Digestibility in humans: [12]

Protein	85%
Fat	91.3%
Carbohydrate	99.5%

In Traditional Chinese Medicine (TCM) the separation between food and medicine is not so sharp as it is in the West. Thus the Chinese eat ginkgo nuts not only for their good taste, but also for their strengthening and tonic properties. In TCM ginkgo nuts are used as a kidney yang tonic, which means that they are capable of increasing sexual energy, stopping bed-wetting and frequent nocturnal emissions, restoring hearing loss and soothing bladder irritation. When boiled as a tea, the nuts are also thought to be a good remedy for lung weakness and congestion— particularly coughs and asthma, because of their strengthening, expectorant, sedative and phlegm-dissipating properties.[13,14] The nuts are especially indicated for wheezing and coughing when there is excess mucous present and, when used on a regular basis, they can also help control vaginal candidiasis ("yeast infection"), frequent urination, cloudy urine and excess mucous in the urinary tract. Other common uses for ginkgo nuts in TCM include eating the cooked nuts as a remedy for intestinal worms, cancer, gonorrhea and

---

"The Chinese eat ginkgo nuts not only for their good taste, but also for their strengthening and tonic properties."

---

leukorrhea, and using them as a poultice for infections (modern studies have demonstrated antibiotic activity).

Although, in TCM ginkgo leaves were used much less than the nuts,[15-21] the Chinese do have some traditional medicinal uses for the leaves. For instance, ginkgo leaves are used for treating chilblains (reddening, swelling and itching of the skin due to frostbite), while a tea of the leaves is sprayed into the throat for asthma.[22] This last use is especially interesting, because modern laboratory studies with humans have found a series of compounds in ginkgo leaves called *ginkgolides* that can reduce the bronchial reactivity of asthmatic patients to common allergens such as dust and pollen. Also, other compounds (called *terpenes*) recently discovered in ginkgo leaves have shown the ability to block the production of *platelet-activating factor* (PAF), which may be one of the primary chemical agents responsible for bronchial reactivity.

Traditionally the Chinese often use ginkgo nuts in combination with other herbs to increase their effectiveness. Three common combinations are shown in Table 2.

TABLE 2

HERBS TO COMBINE WITH GINKGO

The following herbs are all Chinese herbs. You can get them from a local herb store, Chinese herb store and some acupuncturists, or order them by mail from one of the outlets listed in the Resource list (p.74).

CONDITION	FORMULA
Cough, wheezing, and upper respiratory infection with yellow sputum	decoct with *Herba Ephedrae, Semen Pruni Armeniacae*, and *Cortex Mori Albae Radicis*
Vaginal yeast infection	decoct with *Cortex Phellodendri* and *Semen Euryales Ferox*
Frequent urination and incontinence	decoct with *Ootheca Mantidis Fructus Alpiniae Oxyphyllae*

OTHER RECOMMENDATIONS

Average Dose   4.5 - 9 gms/day in decoction

Cautions	Avoid in "excess" conditions, for instance, where there is acute infection with heat. Take for 10 days on, 3 days off, for up to several months. Do not take continuously. Avoid raw ginkgo nuts, which can be toxic.[23]
Tastes	Ginkgo nuts are considered sweet, bitter and astringent in TCM.

Besides ginkgo's use as food and medicine, the Chinese have used the seeds of the plant to wash clothes and, when extracted in wine, to make a preparation that cleanses the skin. Also, the seed's outer coat is caustic and can be used as an insecticide, and ginkgo wood is very beautiful—white or light yellow in color and satiny to the touch—and is used to make toys and chess sets.[24]

MODERN USES OF GINKGO
Important Background Information

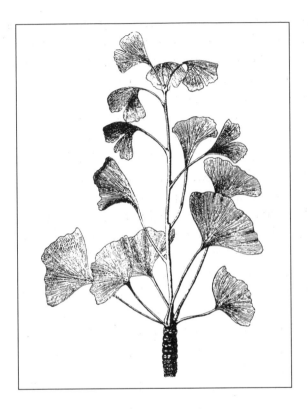

River of Life: The Importance of Proper Blood Circulation

Before we start to explore in detail everything modern research has discovered about what ginkgo can do and how it works, let's stop to review, in simple terms, one of the bodily processes that is essential for good health—blood circulation.

It cannot be overemphasized how vital blood circulation is to our health and well-being. The quality of circulation and how freely blood can move through the vessels to all parts of the body and how well it can carry nutrients (such as oxygen, sugars, enzymes and other life-giving nutrients) and remove the waste products of cellular metabolism directly affects the health of every cell in the body.

When all these nutrients, oxygen, sugars and enzymes are not supplied to the cells of the body continuously throughout the day and night, and the wastes are not constantly removed, something very dramatic happens: all the cells of the body begin to age more rapidly. For when our cells, tissues and organs are not properly nourished and sit in their own wastes, they wear out much faster. We begin to experience aches and pains, we get stiff, our joints start to weaken, and we have trouble running and walking. Our eyes cannot see a beautiful sunset as well—we cannot experience the full range of subtlety in a beautiful piece of music—our senses become dulled. And most profoundly, the brain—that miraculous instrument which allows us to interpret and draw joy and pleasure from everything our senses register—starts to become less alive, less active. It forgets the name of our childhood friend, forgets and doesn't learn new things as well, has trouble keeping up with all the changes going on around us. We start to get old.†

But what causes our circulation to diminish? It can be any number of things. Stress, poor diet, injury and lack of exercise are some common causes.

These days many environmental influences also affect us adversely: radiation from the weakened ozone layer, pollution, pesticides and chemicals, such as lead and other heavy metals in our food, water and air—to name just a few. The sad fact is that too many things in our modern lifestyle are less than ideal for our health.

---

† *What is growing old? Is it how many years we see on our driver's license? Is it the grey hair or wrinkles? Or is it our flexibility? How well are we able to respond to the constantly changing environment? How limber and flexible our mind, body and spirit remain says more than anything how old or young we are, independent of our chronological years.*

For these reasons, a little extra insurance is important if we are to enjoy a vital and healthful old age. There are many ways to provide ourselves with real, natural "health insurance." Certainly keeping emotionally open and sharing the wealth of love we have inside us with others is at the top of the list. Exercise (such as stretching for flexibility), proper nutrition (eating mostly whole, organically-grown foods), being conscious to breathe fully and deeply and preserving a positive outlook on life are other healthy habits to cultivate. Granted, it is often difficult to change habits and thought patterns that no longer work for us, but in the end, it is well worth the effort. For once we have slowly and patiently learned new, more positive habits, we find that it is actually difficult to go back to old, unhealthy ones!

Now one could say, simply, that ginkgo has three main effects on the body:

1. It improves the quality and quantity of the **micro-circulation**, which in turn improves circulation to all the vital tissues and organs, such as the heart and brain. (See page 17.)

2. It helps stop the damage to our organs from potentially dangerous chemicals called **free-radicals**. (See page 39.)

3. It **blocks a common allergic substance** in our body called platelet-activating factor (PAF) which may help cause diseases such as asthma, heart disease, hearing disorders and skin disorders such as psoriasis. (See page 44.)

I will discuss **Free-radicals** and **PAF** later on, so let's take a closer look here at what micro-circulation is. As can be seen in Figure 4, each and every cell in our body is dependent on the "interstitial fluid" which is a sort of "sea" of nutrients and circulating wastes constantly in motion. This life-giving fluid oozes in and out of the tiny capillaries that feed into larger arterioles and venules, then arteries and veins, and ultimately, the lungs, to be oxygenated and to release carbon dioxide. The liver and digestive tract must also supply proteins, carbohydrates, fats, vitamins and minerals.

On the cellular level, millions of chemical reactions and processes are taking place second to second: each cell literally has a life of its own, yet is also part of the larger group of cells which make up a tissue, an organ and finally, the whole body. Blood vessels are the vital "service" lines to the interstitial fluid; they bring sugar and oxygen (the fuel of all life processes); they remove wastes and take them to the liver and kidneys for processing and elimination; and they also bring hormones to speed up and slow down various life processes.

To further emphasize how important this micro-circulation is to good health, in the following sections we will explore in detail the main ways in which ginkgo can maintain and even improve micro-circulation.

One more thing we can do to maintain good blood circulation as we age—and this is the specific subject of this book—is to take ginkgo as a medicine and/or food supplement. Nature has provided us with many herbs and healing plants that have been used since antiquity to help restore health, prevent illness and support and nourish our bodily systems.

Scientific Research and Findings

Despite the fact that the Chinese have traditionally used ginkgo *nuts f*or medicine, most modern research has focused on the *leaves*, because they contain a variety of active compounds. In Europe,

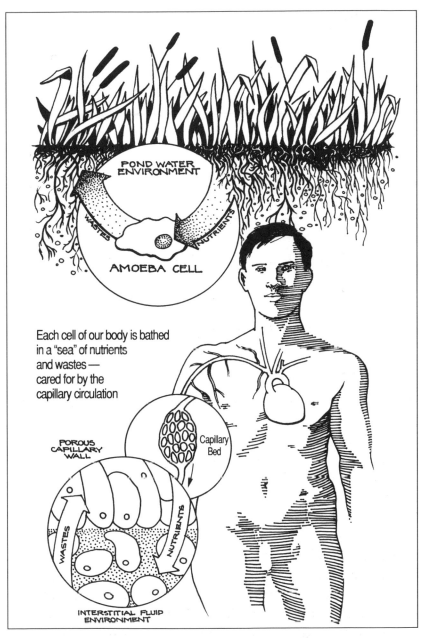

Each cell of our body is bathed
in a "sea" of nutrients
and wastes —
cared for by the
capillary circulation

Figure 4. The proper care and feeding of our cells promotes
health and longevity.

where herbal medicine is more developed and accepted by both the public and health care professionals than in the United States, extracts made from ginkgo leaves are among the best-selling herbal medicinal products. In fact, these products are so popular that recently a European pharmaceutical company started a ginkgo plantation in South Carolina to provide an additional source for leaves to make an extract sold in many countries around the world.[25]

But why, you may be wondering, is there so much interest in ginkgo leaves?

It's due to what the laboratory studies have found—that an

"To maintain good blood circulation as we age... take ginkgo as a medicine and food supplement."

---

extract of ginkgo improves brain functioning in the young and old, prevents or treats circulatory disorders such as stroke and arterial insufficiency (lack of adequate blood supply to the extremities)[26] and can be of use in treating and protecting against hearing disorders, macular degeneration in the eye, and possibly asthma. And this is just the beginning of the findings. During the last fifty years, literally hundreds of scientific studies on a ginkgo extract have been conducted in Germany and France (and a few in Italy, Japan and China), and the physiological effects of ginkgo have been shown to be many and varied. Table 3 summarizes the results of this research and indicates the major types of health problems for which ginkgo has been proven to be an effective medicine. Since most of the scientific literature describing this research is complicated and highly technical, I will not detail it all in this work. However, if you are interested in reading the original research, representative studies are referenced in Table 4. Also, for a complete review of the science, botany, pharmacology, toxicity and pharmacy of ginkgo, see my feature article on the subject in *HerbalGram*.[27]

Regarding Table 4, you will notice that the table is organized into six columns or categories. The first column gives the name of the health problem studied, which may be either the technical name or the name of the disease as it is cited in the original research. The second column gives a description of the condition in more understandable terms. The third column lists the *average* dose of ginkgo extract that was administered during the study, if that information was available. The fourth column specifies the duration of the treatment with ginkgo extract, and the fifth column summarizes the

Ginkgo Improves Mental Health

"Cerebral insufficiency" is a term for a condition that includes a general lack of mental health and vitality, particularly in people over 60. Symptoms of this common disorder of the elderly include decreased memory, intellectual ability, alertness and sociability, along with frequent depression, headache and vertigo.

In a year-long study directed by Sitzer (1987), ginkgo leaf extract proved to be highly effective in treating cerebral insufficiency. A group of 30 patients with symptoms such as headache, vertigo and tinnitus were given a 40 mg dose of ginkgo extract 3 times a day. The researchers found a *"very strong improvement"* in these symptoms after only 2 to 4 months of treatment, and the patients *"kept on improving during the course of the year."* [57]

For cerebral impairment due to degenerative or vascular causes, the average success rate in several open, non-comparative studies was found to be 60 to 78%. In these studies, which had no control group, ginkgo leaf extract was administered orally over periods of 3 weeks to 1 year, at doses of 120 to 360 mg per day.

More carefully controlled, double-blind studies for the same condition confirmed these findings. In a total of 9 studies, which lasted from 5 weeks to 12 months, the overall improvement rate for patients who took ginkgo extract was between 44 and 92%, while people who took placebos showed only a 14 to 44% rate of improvement—quite a significant difference.†

results of the study. The last column provides several representative references for further reading on each type of health problem. Note that all studies were performed with human patients.

In the following sections, I will discuss in greater depth the findings summarized in Table 3, trying always to explain the scientific information in understandable terms. Those interested in more technical information should read the appendix and then look over the reference list for further reading. However, before you delve into this table and the rest of this book, you might also want to acquaint yourself with these basic scientific concepts:

A "double-blind" study is one in which one group of patients is given the medicinal preparation in question, while another group is given a preparation called a placebo (from the Latin, *placeo*, to please) that appears identical to the real thing but is inert. Neither the patients nor the researchers administering the preparations know which group has taken the "drug" and which the placebo until the end of the experiment, when other scientists who were not involved in the testing process evaluate the results. A double-blind test is thus an attempt at reducing the natural tendency of the mind to influence the healing properties of a medicine. Whether or not modern double-blind tests achieve this aim in any significant measure is the subject of much debate, but double-blind tests are still considered to be the most "conclusive."

A special kind of double-blind study is a "cross-over" study. In this kind of study, the patients receiving the placebo and those

---

† *Interestingly enough, these results show not only that ginkgo extracts were helpful in these cases, but also that the attention and care given to the patients in the placebo group made a significant improvement in how they felt. This TLC effect (to give it a quaint name) should be taken as a clear indication of how to improve all treatments and any kind of health care. The very essence of holistic, mind-body healing is the understanding that everything, including human contact and caring, has an effect on health and well-being. Thankfully, today many respected medical journals are beginning to focus on this positive and useful phenomenon, sometimes known as "high-touch" medicine, which I have called the TLC effect.*[54]

receiving the medicine being tested *switch* ("cross-over") treatments half way through the study. The placebo group starts to take the medicine, and the medicine group starts to take the placebo—but always without either the patients or researchers involved being aware of the switch that has happened. Thus a "cross-over" study is another way in which scientists try to strictly control experiments and prevent people's expectations from biasing the results.

Table 3 summarizes the major health problems for which doctors prescribe ginkgo today. As you can see from the table, ginkgo may be the ideal herbal support to counteract some of the most common conditions associated with the aging process and with environmental pollution and stress. This is so because ginkgo, with regular use, can help increase and maintain the blood supply to all the tissues of the body, but especially to the brain, extremities, skin, eyes, inner ear, heart and other vital organs.

TABLE 3

THE MAJOR USES OF GINKGO

In modern clinical studies, ginkgo has been shown to help restore health in the following areas:

✓ **Brain function** Cerebral insufficiency (decreased blood flow to the brain) can adversely affect memory, concentration, intellectual ability, vision, equilibrium and balance. It may also lead to symptoms such as headaches, depression, mental confusion. Stroke can occur because of lack of blood flow and oxygen to brain tissue. Ginkgo can prevent and even treat all of these conditions.

✓ **Circulation disorders** Peripheral vascular disease may cause poor circulation in the legs, making walking difficult. It may also cause poor circulation to the skin, heart and other organs.

✓ **Hearing disorders** Ringing in the ears (tinnitus), disturbance of balance, dizziness (vertigo), sudden hearing loss and hearing weakness may all result from lack of proper blood circulation. These conditions can also be caused by free-radical damage.

✓ **Eye disorders** The retina may be damaged by free-radicals, hemorrhage (as in senile macular degeneration) and perhaps restricted circulation due to stress.

✓ **Senility** Ginkgo protects against brain weakness, improves blood circulation and protects against free-radical damage.

Ginkgo may also be of use in these ways:

✓ **Environmental protection** To guard against tissue and organ damage due to environmental toxins such as pesticides and herbicides.

✓ **Stress support** To improve blood circulation and counteract cellular damage which may arise when excess free-radicals in the body are generated during chronic activation of the immune system.

✓ **PAF-inhibitor** To counteract toxic shock and reduce symptoms of asthma as well as a wide range of other common allergic and immune-based disorders, among them psoriasis and other skin allergies, cirrhosis and cardiovascular disease (stroke and abnormal blood-clotting).

For a detailed discussion of the changes in brain tissues which cause and are caused by "cerebral insufficiency" in the aging brain, see the excellent review by Clostre.[62]

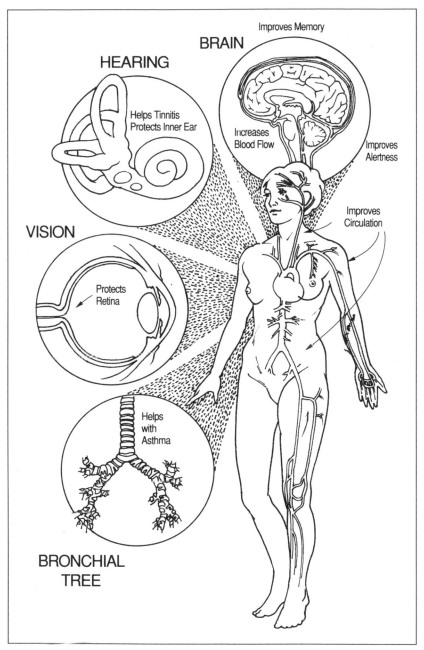

Figure 5. Major Effects of Ginkgo

TABLE 4
THE RESULTS OF MODERN RESEARCH ON GINKGO

Problem	Description	Dose	Duration	Results	Ref.
Circulation and Cardiovascular Effects					
Arterial insufficiency (peripheral arteriopathy)	poor blood circulation to legs, resulting in painful walking	—	65 weeks	Patients who took ginkgo walked further and had less pain than those who took the placebo.	28, 29
Varicose conditions & leg ulcers	varicose veins, loss of veinal tone	—	—	Ginkgo is helpful with varicosities and leg ulcers.	30
Water Retention					
Idiopathic cyclic edema	water and sodium retention in young women due to increased capillary permeability; hormonally related	up to 240 mg/day	1 to 6 months	In a study with 11 women, patients showed significant improvement.	31
Peripheral edema	fluid buildup in the limbs due to increased arterial pressure and capillary permeability	—	—	Peripheral edema was reduced.	32
Mental Effects					
Intellectual function/ cerebral insufficiency	decrease in memory, alertness, attention, mood, awareness	160 mg/ day	12 months	58% of a group of elderly patients who took ginkgo showed improvement, while only 43% in the placebo group improved.	33, 34, 35, 36
				Patients who took ginkgo showed significant improvement over those taking placebo for headaches,	37, 38, 39

				depression, poor concentration and memory, and other symptoms of cerebral insufficiency. Also, improved cerebral circulation was found in patients with Parkinson's disease.	
Memory, short-term	improvement in short-term memory	600 mg/dose	1 hour	Healthy women showed significant improvement in short-term memory over the placebo group.	40
Dementia, cognitive disorders secondary to depression	organic deficiency of the brain related to age	—	—	Mood and cognitive function improved in elderly patients.	41
Hearing Acute cochlear deafness	usually sudden deafness; possibly related to lowered blood flow to cochlea	4 ml 2X/day	30 days	52% average improvement in ginkgo group.	42
Tinnitus	ringing in the ears	4 ml/day in 2 doses	13 months	Ginkgo group improved more than placebo group.	43, 44
Vertigo	vertigo, dizziness (equilibrium disorder)	120 mg/day	12 weeks	Vertigo improved in 20% of placebo group, 50% of ginkgo group.	45, 46
Eyesight Senile macular degeneration	degeneration or hemorrhage in the central area of the retina (a leading cause of poor eyesight in the elderly)	80 mg 2X/day	6 months	Visual acuity improved in ginkgo group only; ginkgo may reduce free-radical damage and the possibility of hemorrhage.	47

BRAIN FUNCTION

One of the most fascinating and encouraging findings about ginkgo leaf extract is that it tends to concentrate in the brain tissues and has a beneficial effect on many aspects of brain functioning, especially symptoms associated with the aging process. For instance, a number of studies have shown that ginkgo significantly improves sociability, alertness, mood, memory, headaches, vertigo and intellectual capacity.[48-53] In one of these studies, a one to three month treatment program using ginkgo leaf extract completely removed symptoms of dizziness, tinnitus (ringing in the ears), headache, disorientation, memory loss and anxiety in up to 92% of the patients, while only 50% got better among the control group of patients who took a placebo.

In another study, ginkgo was proven to increase alertness in the elderly as well as in young people.[55a] Perhaps it seems like no great surprise that the older participants in this study benefitted from the ginkgo leaf extract, but it is intriguing that the 6 healthy younger volunteers also showed improvements in mental alertness. One hour after taking a high dose of ginkgo (up to 600 mg of the standardized extract), their alpha and beta brain waves were stronger, and these effects lasted for 4 1/2 hours.[55b] These changes were tracked with an EEG, or electroencephalograph, which measures minute electrical charges on the scalp that reflect activities of underlying brain cells. The "beta" wave pattern is associated with normal mental activity and concentration, and the "alpha" wave pattern is found to be present when the mind is at rest, and the eyes are closed.

Findings such as these are quite thought-provoking. For the young they may point the way to future improvements in education, for the elderly to improvements in the length and quality of life. Various brain disorders are common in old age, and diseases of the blood vessels (such as arteriosclerosis, or hardening of the arteries) are among the most common serious illnesses of old age.

Also, stroke—which is a loss of oxygen to the brain, usually because of clogged or leaky blood vessels—is currently the number two killer in the United States. As our population ages due to advances in health care, medicine, sanitation and other factors, these problems will become of ever greater concern.

For example, in the 1950s, 8% of the people in industrialized countries (including America) were over 65. In the 1980s, that figure rose to 15%, and it is estimated to increase to over 20% by the year 2025.[56] Because of these demographical changes, medical systems are already strained by the number of aged patients we have today. Thus it seems certain that in the future more emphasis will be placed on preventative medicine—especially when you consider recent reports that even many 30-year-olds show signs of arterial disease.

As the research I have cited demonstrates, ginkgo leaf extracts can play an important role in contributing to a healthy and clear mind in old age—especially if, in the true spirit of holistic medicine, these herbal remedies are combined with a health program that includes exercise and stretching, a diet rich in whole, chemical-free foods and stress-reduction techniques to maintain positive, open thoughts and emotions. I will explain all of these topics in greater depth below as we take a closer look at exactly how the brain works and what factors—including ginkgo extracts—can help maintain a healthy brain.

MAJOR FACTORS IN BRAIN HEALTH

The brain needs two key substances in order to maintain proper functioning: glucose and oxygen. Glucose (a kind of sugar) is a sort of "fuel" or energy source for the brain. The brain needs a steady and constant supply of glucose to sustain all of its activities, especially during times of mental work or stress. Interestingly, ginkgo has been shown to increase the uptake and utilization of glucose by the brain.[58,59]

Oxygen is the other critical brain-nutrient, and unfortunately our modern lifestyle usually doesn't give us the chance to get enough of it. Imagine this typical scenario: you're sitting at a desk in a stuffy office building concentrating on a computer screen. The air quality is poor due to the lack of green plants in your office and the

---

"Try taking an 'oxygen' break in a nearby park."

---

perpetually closed windows. You feel stiff and tired because you have so much to do, and you haven't had any vigorous exercise in weeks (or even months), but you have to push ahead with the work. You become tense and, without realizing it, your breathing becomes shallow.

Question: do you think you're getting enough oxygen? Answer: probably not (unless you're wearing a diving tank!) In fact, even if you work in a nice building by the sea with lots of open windows and green trees outside, you're still not getting all the oxygen your brain could use. Just working at a sedentary job and having shallow breathing all day is enough to create a less than optimum situation. To give your brain more oxygen, try taking an "oxygen break" in a nearby park (you can run, walk quickly or do some other form of exercise). Or, if such exercise is impossible in your situation, at least practice deep breathing for five to ten minutes a day. You may be happy to find how much better you feel—and to know that you're taking important preventive measures against low-oxygen conditions which can contribute to nerve tissue damage in the brain, impaired brain functioning and ultimately stroke.

Thus throughout your life it is of critical importance to maintain a healthy flow of blood which carries glucose and oxygen to the brain. I cannot emphasize this point too much: blood is life! Without an ample supply of blood at every second, the brain—which is the most delicate and complex organ in the body—will immediately suffer adverse effects and perhaps even severe damage. That's why a ginkgo extract, which has been scientifically demonstrated to improve blood circulation and protect brain tissues under conditions of low oxygen,[60] is one of the best-selling medicines in Europe, with over 5.24 million prescriptions written in 1988 alone.[61]

So let's take a careful look now at some things that can harm the brain. As I've explained above, the most important thing for the brain is a constant blood supply. There are a few main ways that this supply can be reduced:

1. Stress, which can constrict blood vessels.
2. Narrowing and damage of the arteries and other blood vessels, due mainly to stress and diet.
3. Abnormal clotting tendency of the blood, which may eventually lead to blocked blood vessels.
4. "Leakiness" of the tiny vessels in the brain tissue itself.
5. Spasms of the arteries.

Stress

Stress is unavoidable and, indeed, in the larger sense of the word, meaning stimulus, stress is vital to our existence. We need constant stimulus in order to exercise all the systems of our body, including the mind. For example, when a baby is born, its immune system is immature and weak; the "stress" of candida and other organisms in the birth-canal are needed to activate the baby's immune system and help teach it to do efficiently the many jobs it will need to do. Likewise the "stress" of roughage and fiber in our digestive tract stimulates the smooth muscles there and keeps them in good tone for processing and assimilation of nutrients. Exercise, too, can stress our skeletal muscles and actually break down tissues, leading to soreness—though the final result may be beneficial, when not carried to extremes. Even the sexual response, which may produce increased heartbeat, blood pressure and deeper breathing is a kind of stress response.

But whether a stress will have a beneficial or detrimental effect on the body ultimately depends on the nature of the stress, its duration and how we perceive it and respond to it. Obviously, most people would find that sexual pleasure had a much happier expression in our body than getting stuck in rush-hour traffic—even though the immediate physical changes in blood pressure, heart rate and breathing are basically the same in both cases. Such is the power of perception and response.

Regarding duration, as Seyle has explained in his famous work on the stress response, *The Stress of Life,*[63] too much stress for too long can be very harmful. By now it is well known that the chronic condition of stress in which the renowned "type A" personality lives (the person who constantly pushes himself or herself to achieve and produce) leads to heart attack and poor cardiovascular health. In fact, I don't think it would be too far-fetched to say that chronic stress can be considered the underlying cause of most diseases. Though it is beyond the scope of this work to discuss this claim in all its details, there are two major ways that stress can profoundly affect our health and well-being.

1. By activating the "flight or fight mechanism" of the sympathetic nervous system, stress takes blood away from our digestive tract and sends it to our muscles, readying us for action. This is why chronic stress usually harms our digestion and assimilation first—and then weakened

digestion and assimilation over a period of
time can lead to immune deficiencies resulting
in cancer, chronic viral infections and a host of
other diseases. It can also lead to a fundamen-
tal reduction in our vitality.

2. By reducing blood circulation to vital organs such
as the brain, kidneys and liver, chronic stress
can deprive these organs of vital nutrients such
as oxygen, sugars, vitamins and other substances that aid the repair of damaged tissues.

The best remedy for the adverse effects of stress is to adopt
more relaxed habits and lifestyles. Svevo Brooks' book *The Art of
Good Living* gives excellent advice in this regard[64] and suggests many
practical ways to cultivate relaxed and healthy habits. From my own
experience I can say that regular exercise, a diet low in processed fats
and rich in whole, fresh foods and practices such as meditation or
other relaxing "letting go" kinds of non-activities, are all important
elements in a lifestyle that reduces stress.

Damage to the arteries and blood vessels

For optimal brain health the arteries and other blood vessels
that supply the brain with blood need to be flexible and free of
obstructing deposits. If there is even the smallest restriction of
blood to the brain, especially in times of increased need such as
during emotional stress or heavy mental work, the quality of brain
functioning decreases radically. And if this restriction of the blood
supply becomes chronic, eventually the brain tissue may be damaged and stop functioning properly. Then come symptoms of
brain impairment such as depression, confusion, headaches, etc.,
and later more serious diseases and conditions such as stroke.
The most important factor that contributes to blood vessel

damage is a high-fat diet, with stress a close second. Heredity also plays an important role, and it is apparent that some people are by heredity much more susceptible to vessel damage due to dietary factors and stress than others.

Diets that have too much fat tend to leave fatty deposits in the arteries and other blood vessels. Over time these deposits can scar, block or reduce the flexibility of the small arteries and capillaries that supply the brain with blood. Thus it is absolutely essential for good

---

"Pritikin was able to completely reverse the damage to his arteries."

---

health to reduce the dietary intake of processed and highly saturated fats such as those found in margarine and in animal meats. The best kinds of fats to eat are ones that are whole and naturally processed without the use of solvents. Polyunsaturated oils, such as safflower and sunflower seed oils, are better than heavily saturated fats such as animal fats and most processed margarines and shortenings.

However, the very best kinds of oils are ones that contain a balance of saturated to unsaturated fats, a good proportion of *monosaturated* fats (oleic acid) and are unprocessed. Or in simple words, I recommend cooking with olive oil and using moderate amounts of lightly-salted unpasteurized butter. Try to get most oils from their original source (whole seeds and nuts), and avoid margarine, rancid fats and oils, shortening and lard like the plague. Also, reduce the use of any kind of fat or oil to a very moderate amount, and combine this low-fat diet with regular exercise—at least 20 minutes of aerobic exercise 4 times a week.

No one has demonstrated the effectiveness of these principles of proper diet better than the famous researcher Nathan Pritikin, who was diagnosed with coronary heart disease and clogged arteries and was given a virtual death sentence. But Pritikin did not just passively accept his sentence. After years of following stringent dietary guidelines and exercising vigorously, he was able to completely reverse the damage to his arteries. Years later, when he died of other causes, the autopsy showed that he had the blood vessels of a 20-year-old!

Abnormal blood-clotting (blockages)

Excessive or abnormal blood-clotting is of course undesirable. The condition called "stroke" often involves small blood clots that either form in or travel to the brain, blocking the blood vessels there and impairing or stopping the supply of oxygen and glucose to the brain. Strokes may be mild or severe, and in severe cases brain cells and tissue are actually killed.

There are a number of factors that contribute to abnormal blood-clotting, the most important ones being heredity, diet and stress. All of these can make blood become more viscous or sticky, which in turn promotes clotting. The tendency to clot can also be provoked by scarred or rough spots on the walls of blood vessels which may damage red blood cells, platelets and other blood cells involved in the clotting process.

Although nothing can be done about heredity (at the moment), diet and stress, as I've pointed out a number of times, can be controlled. To reduce the risk of abnormal blood-clotting one should eat less high-fat and processed foods, exercise regularly and adopt less stressful living habits. Also, ginkgo extract has been shown to prevent the tendency toward abnormal blood-clotting under certain circumstances—specifically, where there is less platelet adhesion and aggregation.[65] Platelets, or thrombocytes, are blood cells that secrete histamine and other chemicals which help regulate clotting and inflammatory processes.

"Leakiness" of the tiny vessels

"Leakiness" of the tiny capillaries in the brain happens when the walls of the blood vessels start to lose their tone. This condition can result from scarring or damage caused by over-activity of the immune system due to stress or a high fat diet. The inflammatory process, which increases secretion of histamine, can be another cause of leakiness. But whatever the cause or combination of causes, leakiness can lead to edema (fluid buildup in brain tissue), which in turn can damage the nerve cells. Decreased capillary tone also causes metabolic wastes to be removed less efficiently and accumulate in the blood. This build-up of wastes can be harmful to any tissues of the body but is especially damaging to nerve tissue.[66]

Fortunately, studies show that ginkgo extract, when taken on a regular basis, can reduce leakiness and increase the blood flow to the brain, protecting against damage to blood vessel walls and increasing the tone of the tiny capillaries that bring blood to the actual brain tissue.[67]

Spasms of the arteries

Spasms in the small arteries in the brain reduce the blood flow to the brain and can be quite harmful. These spasms can be caused by a variety of factors and conditions, but may be mediated by neurotransmitters, the chemical "messengers" that relay nerve pulses from one nerve cell to another. If such spasms occur during conditions of reduced oxygen supply, the negative effects may be compounded.

It is quite significant news, then, that ginkgo extract has demonstrated a relaxing effect on the arteries and arterioles (smaller branches than arteries) which may even be strong enough to counteract vascular spasms.[68] By doing so ginkgo also improves blood circulation. For instance, in another study in which ginkgo extract was given intravenously to 12 patients with cerebrovascular disease, it was found that "cerebral perfusion" (proper infiltration of blood through brain tissue) was increased by over 8%—not only in healthy tissue, but in damaged areas as well. This means that ginkgo did not "steal" blood from damaged areas and cause it to be sent to healthy tissue, but actually increased blood flow in the whole area.[69]

OTHER FACTORS IN BRAIN HEALTH

There are a few final ways ginkgo extracts can help maintain brain health—mostly by protecting against degeneration of nerves and preserving the functioning of neurotransmitters.

Free-radical damage

Nerves are covered with a substance called *myelin* which is vital to preventing nerve impulses from becoming lost, weakened or cross-wired by other neural circuits. Myelin is very rich in unsaturated fats and can be damaged by environmental toxins and molecules called *free-radicals.*

Figure 10. Hydroxy (-OH) radicals can damage nerve cells

The shapely and resilient ginkgo tree is a common sight in cities throughout the world.

PLATE 1

The fruits look like small apricots, but smell like rancid butter. The nut inside is a choice edible food and medicine.

PLATE 2

The male flowers occur on a separate tree from the fruits — they are small, but necessary for pollination.

PLATE 3

In the fall, the bright yellow leaves collect in great numbers under the ginkgo, a fine autumn sight.

PLATE 4

Free-radicals are highly reactive molecules created by our immune system and other cellular processes. They can be generated by the immune system to use as "ammunition" against pathogens (invading organisms) in the body. However, excessive amounts of free-radicals can actually turn around and "attack" healthy tissues and organs. Under "normal" conditions this doesn't happen, because excess free-radicals are "mopped up" by enzymes the body creates specifically for this job.

Unfortunately, though, we no longer live under normal conditions. When things such as saturated and processed fats, radiation due to the weakened ozone layer, herbicides, pesticides and lead and other pollutants from auto exhaust find their way into our bodies, they provoke the production of many more free-radicals. Working and living in low-oxygen conditions also causes the body to produce more free-radicals.[70]

"Free-radicals are a major contributor to the aging process."

This wouldn't be a problem if the body also made more enzymes to mop up all the extra free-radicals; but, unfortunately, it doesn't. In tissues where blood flow is restricted, at the same time that extra free-radicals are generated, the supply of enzymes is restricted—with the result that cellular and tissue damage increases.[71] That's why the issue of free-radicals is becoming a major health concern, not only where the degeneration of the myelin covering on nerves is concerned, but with many other types of tissue as well, such as blood vessel walls and the neurons (nerve cells) themselves.[72] Free-radicals are involved in many, many types of tissue damage, especially where the inflammatory process is present. In fact, there is even a theory (called the "free-radical theory of aging") that free-radicals are a major contributor to the aging process. This theory is now gaining a wide acceptance.[73]

So it comes as something of a blessing to find that ginkgo acts as a powerful anti-oxidant or free-radical scavenger. Like the enzymes the body does not produce enough of at times, ginkgo mops up extra free-radicals and prevents them from damaging the myelin insulation on nerves and other cells in the brain. Ginkgo also prevents free-radical damage to the cells and organs in other parts of the body as well.[74] Indeed, the chemical compounds in ginkgo extracts which are responsible for

Ginkgo Improves Short-term Memory

Two clinical studies show that ginkgo extracts improve short-term memory.

In the first study, which was performed double-blind with 6 human volunteers, it was found that a single 600 mg dose of ginkgo enhanced the speed of information-recall from short-term memory.[77] Doses of 120 or 240 mg of ginkgo extract as well as doses of placebo did not lead to an improvement in short-term memory.

In the second test (also double-blind), 8 healthy female volunteers took various doses of ginkgo extracts or placebo.[78] After giving the volunteers a battery of memory tests, the researchers found that short-term memory was "very significantly improved" *only* in the subjects who took ginkgo. The authors of the study concluded by saying, "*These results differentiate Ginkgo biloba extract from sedative and stimulant drugs and suggest a specific effect on memory processes*". Interestingly, ginkgo extract has not proved to be a general central nervous system stimulant, and neither has it affected the mood of healthy volunteer subjects. Thus ginkgo seems to enhance memory by acting directly on the "central cognitive processes" in the brain.[79]

However, please note that these studies *do not* prove that you should take very high doses of ginkgo extract to increase your memory capacity, but rather that the cumulative effects of a moderate dose taken over a period of time may benefit memory the most.

fighting free-radicals (called *flavonoids*) were shown in laboratory studies to be up to 10 times more potent free-radical scavengers than the flavonoids commonly found in other plant sources, such as citrus peels and many fruits, including blueberries.[75]

Neurotransmitters

Neurotransmitters are chemicals that carry nerve impulses from nerve to nerve. Many neurotransmitters have been identified in the brain and their proper functioning is vital to mental health. As we age, however, one neurotransmitter in particular, acetylcholine, may lose its ability to bind to certain areas of the brain called "muscarinic receptors"[76] —and reduced muscarinic receptor binding has been reported in Alzheimer's patients. Interestingly, in studies conducted with animals, ginkgo extract proved to increase the ability of acetylcholine to bind to important areas of the brain.

PLATELET-ACTIVATING FACTOR (PAF)
A Potent Messenger of Allergies

One of the most interesting aspects of ginkgo's activity is its ability to block a compound called *platelet-activating factor* (PAF). PAF activates several kinds of immune cells (neutrophils, eosinophils, macrophages) and endothelial cells—all of which secrete chemicals that create inflammation and enhance the blood-clotting process. Chronic stressful conditions, including a diet high in processed oils and exposure to environmental allergens or even food allergens (such as wheat for some people), can over-stimulate the production of PAF. When PAF in turn activates too many immune cells, the immune-system can go haywire, producing conditions such as asthma, toxic shock from bacterial sepsis and maybe even atherosclerosis and stroke.

Although some PAF is, of course, essential to the proper functioning of the immune system and is vital in regulating many biochemical processes of the body (such as platelet secretion for initiating the inflammatory response as protection against bacterial invasion), we do not yet understand very well why PAF and other natural "messenger" substances like prostaglandins and histamine can go awry and initiate degenerative or disease-promoting activity in the body. However, we do know that ginkgo extracts hold great promise for protecting against various PAF-induced conditions that arise from over-activation of the immune system or from PAF's influence on other cells it can strongly affect in the lung, heart, smooth muscles (such as in the colon, uterus, and bronchi), kidney, spleen, liver and skin tissue.[80] In the sections below I will explain how ginkgo may help with many of these disorders.

"PAF can go awry and initiate disease-promoting activity"

As an herbalist and holistic health professional, my own feeling is that our modern life with its many common and constant stresses—among them environmental pollution, to name one big one—is changing almost overnight the biochemical processes the human body has developed and adapted to over millions of years of evolution. Unless we are very careful, we could be headed for serious trouble.

Take our modern diet, for instance. For over 400,000 years humans evolved eating plants, grains and animals that had never been touched by pesticides or environmental pollutants, and then in the last 50 years or so we suddenly began to eat predominantly processed-foods stuffed full of preservatives and other synthetic chemicals. Thousands of new chemicals find their way into our diet, air and water every year. We have no idea what effect most of these single chemicals will have on the delicate ecology of our bodies, much less what effect they will have combined.

Nor can scientific research offer us any quick answers or solutions in these matters. Imagine that a substance like PAF is a single note in an intricate series of chemical harmonies playing inside us at every moment. It takes hundreds of thousands of human work-hours and millions of dollars to arrive at just a partial understanding of one musical line (sometimes of even a single note) of this complex harmony! So, in other words, it may be quite some time before we arrive at a complete scientific understanding of substances such as PAF and of herbal medicines such as ginkgo.

Nonetheless, in the meantime we want some practical help and advice right now. Thankfully we do know enough about ginkgo-- both from its long history of traditional use and from modern research—to be able to give some useful indications of how ginkgo works and what conditions to use it for. The sections below describe and explain some PAF-related disorders that may be helped by long-term dietary supplementation with a ginkgo extract.

Asthma and Allergies

The evidence is accumulating that PAF is an important contributor to bronchial asthma. PAF may be released from different immune cells (as well as epithelial cells which compose internal and external surfaces in the body) in response to pollen, house dust and other allergens, and it can cause bronchial constriction, hypersecretion of mucous and inflammatory changes in the bronchial airways which lead to restricted or limited breathing.

Luckily, though, the evidence is also accumulating that a ginkgo extract can help block PAF and fight asthma. In a double-blind trial with 8 asthmatic patients who were given either a standardized mixture of ginkgolides A, B, and C (BN 52063), or a placebo, and were then asked to inhale allergens using an atomizer, the researchers reported that patients who took ginkgo had a *"significantly inhibited response to the allergens"*.[81] In other words, these ginkgolides may reduce an asthmatic's bronchial constriction in response to house dust, pollen and other allergens—which means that ginkgo extracts may help reduce the frequency or at least the severity of asthmatic attacks. However, actual clinical usefulness of BN 52063 has yet to be proven.

Heart Disease, Stroke and other Circulatory Disorders

When human blood-cells called *platelets* (which help promote blood-clotting) encounter PAF, they change shape and begin to stick together. It takes only a minute quantity of PAF to cause platelets to bind irreversibly. Of course platelets keep us alive by stopping bleeding in emergencies, but when they stick together or "clump" on a chronic basis, they can make blood more viscous and may also secrete chemicals that cause blood vessels to constrict and help form plaque deposits which scar and narrow blood vessel walls. Thus when there are excessive levels of PAF in the blood, platelets may create conditions which lead to heart disease, stroke and other circulatory disorders.

"Ginkgo extracts can help block PAF and fight asthma."

In extensive tests with human blood, it has been found that the ginkgolides found in a ginkgo extract specifically prevent PAF from binding to platelets and causing the whole range of undesirable effects mentioned above, including arterial thrombosis[82] and spasm of the coronary artery (which results in lack of sufficient blood to the heart muscle).[83] Several studies have shown that the heart may be strongly affected in the process of anaphylaxis (an exaggerated immune-mediated reaction to protein), by inducing arrhythmias and reducing coronary blood flow and contractile force.[84] Ginkgolide B shows the ability to prevent damage in this reaction.[85]

Peripheral Arterial Insufficiency

Restriction of blood flow to the legs and other extremities--peripheral arterial insufficiency as doctors call it—becomes increasingly common with age. Usually it is due to hardening of the arteries, plaque build-up, loss of blood-vessel flexibility and other consequences of poor living habits such as smoking cigarettes and eating foods high in processed and saturated fats. Restriction of blood flow in the legs, in particular, can be a troublesome and frustrating condition, for it limits the distance a person can walk without experiencing severe pain. Thus those who suffer or know someone who suffers from this disorder will be happy to learn that it can be treated.

In my experience as a health professional, I have found that removing the risk factors that are associated with peripheral arterial insufficiency in general is the first and most important step in starting to heal this disease. I knew one 68-year old man who was diagnosed with peripheral vascular disease. He had trouble walking for more than a short distance without having severe pain in his hips and upper legs. He smoked, drank coffee and ate a diet rich in animal fat. After changing to a strict macrobiotic diet that emphasized whole grains, miso soup, sea vegetables and a sharp reduction in animal fats and coffee, within several months he was able to walk completely pain-free.

Quitting smoking has many other benefits as well, but in the case of peripheral vascular disease it is almost essential to complete success; I can recommend this first step with no reservations. Regarding reducing the intake of animal fats, I recommend doing this as much as possible, but it is not necessary to be totally vegetarian. Vegetarianism certainly has its own benefits for people with suitable constitutional types, but people who don't feel disposed to becoming vegetarian can do very well by eating fish and chicken in place of other meats, which generally contain much more fat. In any case, however, it is good to focus the diet more on fresh, whole foods such as lightly steamed vegetables, whole grains and legumes and some fresh fruit (see the reference list for cookbooks).

If you have developed the habit of eating fatty snacks, try replacing them one at a time with more healthful ones such as carrot and celery sticks, whole grain crackers and lightly-salted tortilla chips made with unsaturated fat. Occasionally even ice cream and other treats can be beneficial because of their nurturing qualities—though only once in a while. This also depends on a person's constitution. For some people who have severe heart disease, discipline is necessary to choose only the highest quality fresh food, with very few exceptions. In any case, though, remember that it is always best not to force ourselves to change our entire diet all at once; there's no need to shock your body. We must be gentle with ourselves: one step at a time is best except in all but the most extreme cases.

You probably won't be surprised to read (given the subject of this book!) that ginkgo extract has been studied for its effect on peripheral arterial insufficiency in a number of clinical situations and has proven to be quite effective in increasing the distance sufferers of arterial insufficiency can walk without feeling pain. In one of the better-designed studies, with 79 patients diagnosed with arteriopathy (a diseased state of the arteries) in the lower limbs, half of the group was given ginkgo extract (40 mg dose 3x/day), while the other half was given a placebo. Using both objective and subjective measures of improvement, a highly significant increase in the ability to walk for distance without pain was found in the ginkgo group only.[86] In this study and others, ginkgo extract also reduced cramping and numbness of the extremities.

These studies have opened the door to using ginkgo extract as both a preventative and a supportive therapy for many kinds of arterial disease, including arterial insufficiency of the legs.

Hearing Disorders

The major causes of hearing disorders are reduction of circulation in the ears due to loud noises or other causes (such as athero-

---

"Besides hearing problems, ginkgo can also be helpful for dizziness, vertigo and other equilibrium problems."

---

sclerosis); brain insufficiency; and cervical syndrome resulting from strain or injury to the neck. With these types of disorders, too, ginkgo has proved to be helpful. For example, one study combined chiropractic manipulation with regular doses of ginkgo extract, and it was found that this combination was more effective in restoring hearing than chiropractic manipulation alone.[87] In another study, hearing weakness due to inner ear problems was treated with ginkgo extract for a period of 9 weeks. The results showed that 35 of the 59 patients in the study had either "successful" or "very successful" improvements in hearing.[88] Similarly, when 350 patients with hearing defects due to old age were treated with ginkgo extract, the success rate was 82%. Furthermore, a follow-up study of 137 of the original group of elderly patients 5 years later revealed that 67% still had better hearing.[89]

Tinnitus (ringing in the ears) has been observed clinically since antiquity. Three factors are important in determining how successful treatment for tinnitus will be — whether the ringing is constant or intermittent; whether it is in one or both ears; and whether the ringing has been present for more than a year. If the ringing has been constant and in both ears for more than a year, the chances for a successful treatment are greatly reduced.

Several tests using ginkgo extract to treat people with tinnitus of varying degrees of severity have shown that this ailment, which is usually difficult to treat, yields nicely to ginkgo extract. For example, in 1986 a study statistically proved the effectiveness of treatment with ginkgo extract for tinnitus: the ringing completely disappeared in 35% of the patients tested, with distinct improvement seen in as little as 70 days![90]

Besides hearing problems, ginkgo can also be helpful for dizziness, vertigo and other equilibrium problems associated with reduced circulation in the inner ear. The combined results of 5 double-blind studies with ginkgo extract demonstrated that a daily dose of 60 to 160 mg was effective for the above symptoms in 40 to 80% of patients—which was a significantly better success rate than with placebo.[91]

Eye Disorders

The compounds in ginkgo extract has been shown to concentrate in the eye.[92] This combined with ginkgo's PAF-inhibition effects, free-radical and anti-oxidant properties and its strengthening activity on the small capillaries, could lead one to suppose that ginkgo might help protect and restore healthier functioning in the eye. And indeed this is the case. Studies with animals show that ginkgo extract can offer protection in the retina[93] and cornea. The cornea is the most exposed part of the eye and is open to injury and infection. One of the ginkgolide compounds in ginkgo — ginkgolide B (BN 52021)—has been shown to inhibit PAF-mediated inflammation in the cornea, when it is applied topically on the cornea itself.[94]

Senile macular degeneration, which is associated with hemorrhage, pigmentation disturbance and scarring of the macular (central) region of the retina, is a leading cause of vision loss in the elderly. Unfortunately, no effective treatment for this ailment is currently known. We do not yet understand exactly what causes macular degeneration, but it may be the result of an imbalance in which the immune system sends out antibodies against itself to the macular region of the eye. This may provoke tiny hemorrhages in the retina, which in turn can lead to the formation of a scar and loss of sight in the affected part of the eye. However, the disease process is a complex one, and high light intensity (and thus heat), as well as free-radical damage, may also play a role.

To see if ginkgo extract could benefit macular degeneration, a small double-blind clinical study was conducted with 20 patients over 55 years of age who were diagnosed with senile maculopathy.[95] Ten of the patients received 80 mg of ginkgo extract a day for 6 months, while the other 10 received a placebo. After 6 months of treatment, the ginkgo group showed a "significant improvement" in acuity of distance vision. The authors of this study concluded that much more work needs to be done to firmly support the use of ginkgo in this common eye disorder, but the initial results offer great hope. Thus ginkgo is well worth a try, at least as an adjunct therapy, in conditions where the function and integrity of the retina might be compromised, such as in diabetes or glaucoma.

SUMMARY OF GINKGO'S EFFECTS ON THE BODY

By now you've probably read more than you ever wanted to know about ginkgo! Hopefully the scientific details have not been too confusing, and you have emerged with a more precise idea of how ginkgo affects the body. To refresh your memory about

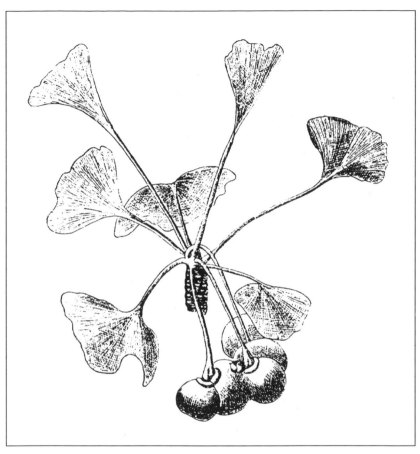

Figure 11

ginkgo (bad joke!) and pull all this new information together, take a look at Table 5, which summarizes ginkgo's major effects on the body. Again, for a more thorough discussion of these matters, see the appendix and the list of references for further study.

As a closing comment, you might be interested to know that there is some evidence that ginkgo may be useful as a topical cream or orally for inflammatory conditions such as sunburn, eczema, acne, psoriasis, rashes and skin allergies, and as a spray for hay fever or inflammation of the sinus cavity.[96, 97, 98]

TABLE 5

SUMMARY OF GINKGO'S EFFECTS ON THE BODY

- **Improves the characteristics and condition of the blood.** Lowers blood's viscosity and platelet adhesiveness; protects red blood cells by stabilizing their membranes; increases blood vessel tone; stabilizes capillary permeability.[99]
- **Free-radical scavenger.** Protects cell membranes in the brain and other tissues throughout the body against free-radical damage.[100] Cell membranes are particularly sensitive to free-radical damage, which can lead to destruction of the entire cell.
- **Increases the uptake and utilization of oxygen and glucose in tissues throughout the body.**[101]
- **Increases blood flow to the brain or extremities.**[102]
- **Regulates or increases brain metabolism.** This counteracts depression.[103]
- **Regulates neurotransmitters.** This helps against memory loss, depression and senility.[104]
- **Protects myelin.** Protects the covering on nerves against some kinds of damage.[105]
- **Protects hearing.** Helps protect against and even restore impaired hearing, especially when due to damage from loud noise or infection.[106]
- **Shows anti-bacterial and anti-candida activity.**[107]

MEDICINAL PREPARATIONS

Now for some very practical and important information—how to actually *use* ginkgo as a medicine. There are two main ways to take ginkgo. First, you can grow your own trees, or harvest the leaves from trees growing away from roads and sources of pollution and make your own tea or liquid extract. Second, there are several different kinds of commercial ginkgo products available in natural food stores throughout the country. If you are interested in the dried seeds of ginkgo, they are available under the name *Pak-Ko* from most Chinese herb stores, acupuncturists or Chinese herb distributors in the United States. These seeds make a delicious addition to rice cereal, soups and other dishes (see the recipe in the section on Traditional Uses of Ginkgo).

Tea

Tea is a traditional way to take ginkgo and when used over a longer period of time (up to 9 months), it may provide good benefits, especially as a preventative supplement to the diet. A tea of the leaves can be a good way of taking ginkgo because most of the important active compounds in the live plant are water-soluble. Because of changes in the levels of active constituents during the growing season and variations in levels of these compounds between different populations of ginkgo, the use of the leaves is best as long-term therapy, mostly as prevention. One drawback of the tea, however, is that it has a bitter, astringent and slightly sour flavor—which is fine for an herbalist who actually *likes* bitter herbs but not for the average person. Nonetheless, if you're interested in tea, don't give up hope yet. I've found the following formula to be quite palatable:

Brain and Circulation Tea

ginkgo	1 part
lavender	1/2 part
wood betony	1 part
lemon balm	1 part
stevia herb	1/8 to 1/4 part

Adjust the amount of stevia according to how sweet you want your tea. Since stevia is very sweet, you might start on the low side.

To round out the flavor of this formula, I add a little lemon juice and a sweetener, such as honey, fructose or sucanat (granulated cane juice). Put 1 oz of the mixture of these herbs in a pot (glass or stainless steel, never aluminum), cover it, and simmer gently for about 5 minutes. Then let the brew steep for about 10 minutes before tasting. Drink one cup morning and evening, or more if you enjoy it. The herbs in the above formula (except stevia) are traditionally used to enhance mood and heal the head. Gotu kola (*Centella asiatica*), the famous Ayurvedic herb, can also be added to enhance brain function, according to its traditional use. I would suggest a fresh-plant liquid extract of the leaves of gotu kola, because the herb loses its properties very rapidly after being dried. It is also possible to grow gotu kola in a terrarium as a house plant. The live plants are available from Forest Farm (see the Resources list at the end of this book).

Harvesting the leaves

The best time to harvest the leaves for high flavonoid content is in the fall, after they begin to turn color. Flavonoids, which may be the most active anti-oxidants in ginkgo and exhibit a vessel-toning effect, reach their highest levels in the leaves at this time.

However, the most important active constituents in ginkgo leaves (the ginkgolides and bilobalide) reach their highest levels around August or September, in late summer or early fall, before the leaves start to turn color. If you are interested in getting maximum levels of these constituents, you should harvest the leaves when they are a dark, rich green, because these constituents drop to their lowest levels after the leaves have turned yellow and begin to fall off the tree.

Considering all of the above, the optimum time for harvesting ginkgo leaves may be just about the time they begin to turn color, or a little before in September.

After picking the leaves, shade-dry them on a screen, allowing good air circulation from the bottom and the top. The best condition is one that will dry the leaves quickly (in 2 to 3 days) but not overheat them. Never dry the leaves in the sun. To test that the leaves are thoroughly dried, "snap" a leaf stem to make sure. When the leaves are dried, put them in a paper bag inside a plastic bag, and store them in a cool, dark place—in a tinted glass jar is best.

Commercial Products

There are several kinds of commercial products available where natural products are sold.

- Powdered leaves in capsules or tablets
- Liquid extracts, also called "tinctures"
- Powdered extracts (usually 6:1 or 8:1), which are unstandardized
- Standardized extracts, 24% flavone glycosides

Which are best? Powdered leaves in capsules are probably not concentrated enough to have much effect and are not really worth using. Liquid extracts of ginkgo leaves, however, can be very good. These preparations are made by soaking the fresh or dried-and-powdered leaves in a solution of alcohol and water for a period of at least two weeks. The alcohol acts as an excellent solvent, pulling the active compounds out of the leaves. Then the resulting "tincture" or liquid extract is pressed out and filtered.

Most of the commercial preparations of ginkgo are standardized extracts. A concentrated powdered extract is the preferred dose form in Europe and Japan. The best-selling products, *Rökan* ® and *Tanakan*®, are standardized to a 24% content of flavone glycosides, but other important constituents, such as bilobalide, are also quantified. All of the studies performed on ginkgo have used this standardized extract, so its effectiveness has more credibility than other preparations. I have also seen ginkgo extracts standardized to 10% flavonoids. These latter products may well be effective, but they have not yet been supported by actual clinical trials.

The German company Schwabe was the main force behind the development and testing of ginkgo preparations for many of the conditions for which we use ginkgo today. Now there are many other ginkgo extracts coming from Europe and Asia beside the original Schwabe preparation—and these other extracts are sold in a variety of products in the United States, Europe and other industrialized countries—but most of these extracts are still standardized to the 24% content of flavone glycosides established by the original Schwabe product even though their ratios of flavonoid, ginkgolide and bilobalide content vary. Achieving the ideal balance of these important constituents is thought by Schwabe to be essential in assuring the effectiveness of the final extract.[108]

Standardization

At present, there seem to be two sides to the debate over whether or not to standardize extracts. One side claims that an herbal product should have a consistent and high concentration of active constituents, so that every time a dose of ginkgo extract is taken, it can be assumed that the effect will be the same or similar to that of previous doses. Standardization also insures the proper identification of plants used in a natural medicine, because the plant's unique constituents or combination of constituents are identified in every batch. When a plant medicine is standardized, but also highly purified, in order to bring the levels of active constituents to a high level, it becomes more like the synthetic drugs used in hospitals and prescribed by doctors—though the herbal product is still made from a natural source.

The other side in this dispute argues that it is better to make a ginkgo extract using only alcohol and then confirming the active constituents to be present at a certain level or even leaving them unstandardized. Some herbalists (traditional purists) feel that this method, because it does not interfere with the internal balance of the herb, is a more natural one. They point out that when constituents are "pumped up" from their naturally-occurring levels, one might lose the total effect of the whole herb. Also, it is said that purified extracts promote the use of herbs allopathically to treat symptoms, not as a whole-plant preparation used in a traditional or holistic way.

Personally, I believe that there is a need and use for both kinds of preparations, according to the situation. Many herbalists would agree that a highly purified plant extract is better than a synthetic drug in certain situations—for instance, in a case of severe mental impairment due to aging which is aggravated by a lifetime of moderately unhealthy living habits (which is the norm in this country). In this case the extra power and consistency of a purified product may be needed. On the other hand, for prevention and as a nutritive daily tonic, perhaps the more slow-acting, whole-plant extract would suffice. Of course, in this regard supporters of standardized extracts might point out that unstandardized products do not have clinical testing and thus can't be counted on—to which traditional purists might reply that ginkgo leaf has been successfully used as an unstandardized preparation in China for thousands of years as an asthma remedy and brain tonic.[109]

My goal in this work is not to argue for or against either side of this issue, but to present both sides and let the reader draw her or his own conclusions. Table 6 summarizes the pros and cons of the main ginkgo preparations.

TABLE 6

PROS AND CONS OF
VARIOUS GINKGO PREPARATIONS

Preparation	Description	Positive Points	Possible Drawbacks
Tea	whole leaf	inexpensive; whole plant in traditional form; may be suitable for self-care and prevention where no serious illness exists	activity may vary from batch to batch; whole leaves are not always available where natural products are sold and must often be self-picked
Tincture	alcoholic extract	moderately priced; plant in whole form suitable for self care and prevention where no serious illness exists	activity may vary
Standardized extract (Schwabe product)	original 24% flavonoids	consistent quality and effect; many clinical trials prove effectiveness— the "original" product	highly purified; loses original internal balance of the herb, according to herbalists who adopt a 'classical' attitude; according to Schwabe, unfavorable constituents are removed, and a good balance of favorable ones is retained
Standardized extract	copy of original extract at lower concentration	more consistent effect than tea or other unstandardized products; levels of flavonoids guaranteed; initial cost is lower than Schwabe product	no actual clinical studies on exact product to support claims; balance of active constituents are different than original Schwabe product

Which Product is Best for Me?

As discussed above and shown in Table 6, there are several ways of looking at what is the most desirable ginkgo preparation. If you are a person who believes that a whole plant extract, though less potent, is the natural and therefore the best form of ginkgo for you, then you should choose to make your own tea or buy a liquid extract product. A liquid extract may be more suitable for prevention or rebalancing of minor health problems. There are several companies in this country, and one respected Swiss company, making liquid extracts from "wildcrafted" or organically-grown leaves harvested from trees that grow away from roads and sources of pollution. These products are desirable because of their makers' commitment to the organic movement and high standards of quality, though they come with no assurance of standardization.

On the other hand, if you are a doctor or practitioner, or a person who appreciates a well-researched, purified product, then you will probably prefer the original Schwabe product, available in France under the name *Rökan®*, and in Germany under the name *Tebonin®*. This product is also available in the United States (see Resource Section).

Lastly, if you want the consistency of a standardized purified product but don't consider the extensive research behind the Schwabe product a deciding factor, then you can choose one of the many other standardized extract products available, some of which are good quality. Though the Schwabe product usually has the highest initial price, a Schwabe representative has told me that because of their many years of research, and the perfect balance of all active constituents in their extract, they feel their product is more effective, and thus a good value in the long run.

Dosage Information

There are good reasons to believe that ginkgo will be most effective when taken over a period of time. Studies on absorption show that half of the active compounds in ginkgo may be eliminated from the body after 3 hours and the rest within 20 hours. This elimination by the urine is almost complete, for only 2% of the active compounds may remain in the body over the long-term.

Interestingly, 72 hours after administration of ginkgo, higher concentrations of the remaining active compounds were found in the hippocampus, striatum, hypothalamus, eye lens, thyroid and adrenals than in the blood. The significance of this "selective" concentration of ginkgo compounds is not yet known, but may partly explain why organs such as the eyes, ears and parts of the brain are predominantly affected.[110]

The bioavailability, or overall ability of the body to absorb and utilize ginkgo's flavonoids, was determined to be around 60% in rats. In humans, clinical studies show that the standardized extract of ginkgo is well-absorbed, a maximum absorption being reached after only 1 hour.[111]

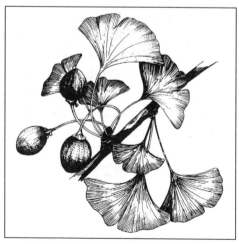

Figure 12

TABLE 7

DOSAGE INFORMATION FOR GINKGO

Clinical studies show that ginkgo is effective after a few days at a high dose or after 1 to 9 months at a lower dose. Dosage information is summarized below.

Preparing a Tea

Put 1 ounce of the dried leaves into 1 pint of water. Simmer over low heat for 5 minutes, let cool, and drink a quarter cup warm, morning and evening. Add sweetener or other aromatic herbs (such as chamomile or peppermint) to taste.

Taking a Liquid Extract

Put 40 drops in a little water or juice and drink morning and evening.

Recommended Dose for Standardized Extracts (24%)

A therapeutic dose of the Schwabe tablets is 1 tablet 3 times daily, where one tablet contains 40 mg of the 50:1 standardized extract equal to 9.6 mg of ginkgo flavone glycosides. For peripheral vascular disease, tinnitus and vertigo, a higher dose may be needed (160 mg/day, given as 80 mg twice daily, or 40 mg 4 times daily).[112]

SAFETY AND TOXICITY

The Leaves

Ginkgo leaves and cooked nuts have a safe history of use that goes back several thousand years. For this reason, when using a preparation of the whole leaves or seeds, you can safely assume that within a normal dose range there will be no danger of negative side effects.

Because the highly purified extract (24%) is so concentrated, it has a stronger activity per amount of extract than the ginkgo preparations used for thousands of years in Chinese medicine. Thus not only has the internal balance changed in the purified extract, but so has the potential toxicity, both short-term and long-term. Fortunately, Schwabe has done an enormous amount of testing which demonstrates that the purified and concentrated extract is safe. Thousands of people have taken high doses of the extract and have been rigorously examined for any undesirable bodily changes.

"Within a normal dose range there will be no danger of negative side effects."

Nonetheless, when the dose is high enough, or if the extract is injected into the body rather than taken orally, there have been a few minor side effects reported. For example, in one study involving 2,855 patients who took ginkgo extract, about 3.7% experienced minor gastric upset which had no lasting effects when the ginkgo was discontinued.[113] Another test with 8,505 patients who took ginkgo for 6 months revealed that only 0.4% (33 people) experienced minor side effects, most commonly mild stomach upset.

It is noteworthy that tests show that even high doses of ginkgo extract do not change the hormonal balance in men,[114] and that neither does ginkgo affect the sugar metabolism of the body—which means it is safe for diabetics, who often suffer from poor circulation and therefore might benefit from ginkgo treatment.[115] Finally, no disturbances in the formation of new blood cells or the functioning of the liver and kidneys were observed, even after long-term use.[116]

The Fruits

Ginkgo fruits have been the object of much bad press—due to their potentially foul odor, which smells something like rancid butter. Because even a few ginkgo trees can generally disgrace the sidewalks of an otherwise respectable neighborhood, in cities the pollen-bearing trees are usually planted instead of the fruit-bearing trees.

Aside from its unaesthetic odor, the pulp of ginkgo can actually be irritating to the skin and cause rashes in some people, for it contains similar irritating chemical substances to those found in poison oak and poison ivy. Many cases of contact dermatitis have been reported from handling the fruit when its skin is cut or broken,[117,118] though rashes have not been reported from contact with the leaves or any other part of the tree.[119]

In China, traditional processing methods are used to remove the pulp cleanly from the nuts.

KEYS TO GETTING WELL

Often the questions most on the mind of a person who wants to restore lost health are how soon will I begin to feel the results, and what will I feel? Is this really going to help me?

These are good and understandable questions, and my answer is Yes, ginkgo can help greatly, but it often takes patience and perseverance—and you may have to try not just one but several remedies before you find the single herb or combination of herbs that works for you. When we are sick enough to lose our sense of well-being, or if an ailment is chronic, we may begin to feel as if we will never get well. That's why it is a very positive step to seek help in as many ways as possible. We must keep searching for the answer to our particular health problem until we find the answer. It is often useful to read books about health, especially inspiring, positive ones that can give us hope and help restore an optimistic outlook (see the reference section for some that I have found particularly inspiring in my own healing process). Also, we may need to try several different natural remedies derived from plants. For the most part, herbal medicines are mild enough so they don't place an extra burden on us when we are already weighed down with worry and illness.

"How a medicine will work with an individual depends not only on the medicine and the dose but also on how each person responds to the medicine."

For many of the diseases mentioned in this book, ginkgo is certainly well worth a try. It has many beneficial and protective effects and in all likelihood will cause no side effects. Of course, each person is different. Each of us has a different constitution and nature. Traditional systems of healing such as Traditional Chinese Medicine (TCM) and Ayurveda have paid particular attention to these differences, and practitioners of these healing systems will prescribe an herbal remedy based on what is called a person's "constitution". For instance, one person may have very weak digestion and poor recuperative powers while another may be very robust and have powerful digestion. Thus the latter person can take a stronger medicine, while the former one may need "building up" before he or she can take a stronger metabolic-activating preparation. This precisely is why some people may experience digestive upset with ginkgo while others may not.

How a medicine will work with an individual depends not only on the medicine and the dose but also on how each person responds to the medicine. Every substance, whether food or medicine, has the potential to either support the body's healing process or harm the body. That's why both the medicine and the person have to be considered for herbal medicine to be completely effective.

One way to help counteract individual differences in response to a purified ginkgo product is to combine it with other herbs (such as ginger) to help warm the digestion. In TCM, ginkgo nuts are often mixed with as many as 10 or 12 different herbs, depending on the constitutional type of the person who will be taking the preparation. Also, it is possible that ginkgo just is not the right herb for a particular person. No herb or medicine is right for every person at every time.

Nonetheless, there are some herbs that are almost universal as protecting and balancing herbs—and ginkgo seems to be one of them. This is said with the understanding that a highly purified preparation may not be indicated for long-term, daily use as a "food" by healthy individuals. For them a milder ginkgo preparation would be recommended, or a combination of building and nutritive herbs such as astragalus, eleuthero, unprocessed ginseng, reishi and shiitake. However, considering that circulatory diseases, environmental stress (such as industrial pollutants) and immune imbalances are unbelievably common today in developed countries throughout the world, this still leaves a wide scope of useful application for standardized ginkgo preparations.

APPENDIX

BOTANY

Ginkgo biloba is the sole representative of the genus Ginkgo, from the family Ginkgoaceae and the order Ginkgoales, from the class Gymnospermae.[120] In other words, ginkgo is more closely related to pines and other conifers than any of the angiosperm trees such as maples and oaks. It is separated from the other families in the class by having flattened leaves and motile male sperms.

The ginkgo is what is called dioecious, "two houses," because the male flowers (slender, stalked catkins) are borne on one tree, while the female fruits on another. Three to five leaves grow from a thickened stalk, are wedge-shaped and have many parallel veins. The leaves turn yellow in the fall and eventually drop.[121]

When young, the ginkgo tree grows slowly and for the first 20 years or so is a slender tree with a height of up to 20 to 30 feet. Later, as the tree matures, the lower limbs become heavier and fuller and more spreading. I have sat in the branches of a ginkgo tree that was over 100 years old. The tree was massive and quite gnarled, reaching up nearly 80 feet and spreading for 20 feet in every direction. As they mature, ginkgo trees take on a magnificent visage, seemingly timeless in their statuesque beauty.

There are several different forms of ginkgo known in cultivation:[122]

1. var. *fastigiata*, which is columnar in form and has almost erect branches
2. var. *macrophylla laciniata*, whose leaves are larger and more deeply cut than in the type form
3. var. *pendula*, with branches that are more or less weeping
4. var. *variegata*, whose leaves are variegated with pale yellow streaks

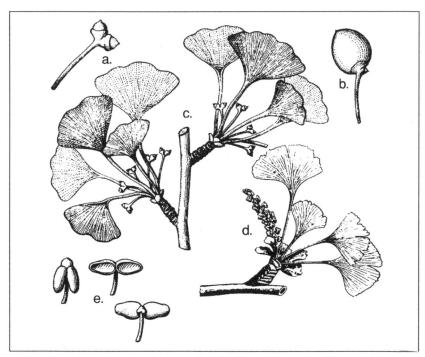

a. Fertilized ovary   b. Immature fruit   c. Fruiting branch
d. Male inflorescence   e. Pollen – bearing anthers

Figure13. Details of reproductive structures

CULTIVATION

Cultivation of ginkgo is not difficult, but it is a slow-growing
tree, and sometimes it takes a little energy and care to get ginkgo
trees established. In the eastern United States, ginkgo does well on
rich soil as far north as eastern Massachusetts and central Michigan.
Along the St. Lawrence River, it grows well up into Canada.[123] In
many other areas of the country it is commonly grown for its
upright stature, beautiful leaves and hardy resistance to city conditions, pollution and attacks from insects, viruses, fungi and other
pathogens. It is thought that ginkgo's remarkable stamina and
immunity is due, at least in part, to the constituent 2-hexenal.[124]

Ginkgo can be grown from seed, stratified in the autumn. Place the seeds in moist sand in a plastic bag for several months in the refrigerator. In the spring, take them out and plant them in a good potting mixture. Many horticultural varieties have been developed, some with variegated or fringed leaves. Ample nitrogen and optimum soil conditions will, of course, enhance the growth.

Ginkgo trees do not flower or grow reproductive structures for at least 20 years, but they can continue to reproduce in some cases for 1,000 years. Though one only needs a few trees for personal use (such as for making tea), patience is required while growing them. After about 10 to 20 years, they become large enough to really start producing. I have watched a 5-year-old tree for several years and so far have seen 12 to 18 inches of new growth per year. However, my 1-year-old has nearly doubled in size this last year. It is possible to harvest a crop of leaves for commerical applications after a year or two, but a large number of trees is required. Fortunately, harvesting the leaves places only minimal stress on the tree because the ginkgo will shed them in the fall anyway.

CHEMISTRY

The leaves of *Ginkgo biloba* exhibit a complex array of secondary metabolites, listed below.[125, 126]

1. Several groups of flavonoids
    a. Heteroside flavonoids, the most abundant aglycones which are quercetin, kaempferol, and isorhamnetin. The commercial European and Asian extracts are usually standardized at 24% of these.
    b. Coumarin esters of quercetin and kaempferol and glucorhamnoside
    c. Proanthocyanidins (ionized flavonoids), in condensed form (dimers and polymers), chiefly from delphinidin and cyanidin
    d. Catechins

e. Proanthocyanidins (gallochatechin, 4,8'' -catechin (=procyanidin), etc.

f. Condensed tannins

2. Terpenoid substances (which impart a bitter flavor to ginkgo)

    a. Several unique diterpenes, namely, the ginkgolides A, B, C, J, M. Clinically, these show the most promise. They occur in the leaves at about 0.2%, but the original extract contains about 6%. They were synthesized for the first time by E.J. Corey and colleagues at Harvard University in the first months of 1988.[127]

    b. One sesquiterpene—bilobalide

    c. The sesquiterpene biflavones—sciadopitysin, ginkgetin, isoginkgetin and bilobetin (also called bilobalides GKA, GKB, GKC)

3. Organic acids—hydroxykinurenic, kinurenic, protocatechin, parahydroxybenzoic, vanillic acid, and ascorbic acid (vit. C). These substances add an acidic character to ginkgo extracts and render them more water-soluble.

4. Carotenoids, such as lutein epoxide, zeaxanthin, and others

5. Iron-based superoxide dismutase (SOD)

6. Sterols, such as sitosterol

    Probably the most interesting and potentially useful of the many different kinds of compounds found in ginkgo are the terpenoid compounds called ginkgolides. These compounds are unique to ginkgo and are found in the leaves in small concentrations (about .2%) and in the largest concentrations in the root bark (about 1%). Unfortunately, harvesting the roots, which would afford a rich supply of these important active compounds, would destroy the tree. Ginkgo is a slow-growing tree, and therefore the leaves are the best part to harvest, being a renewable resource. It is interesting that the ginkgolides are very difficult for the chemist to synthesize, though the ginkgo tree seems to do it easily.

	R	R'	R"
Ginkgolide A	OH	H	H
Ginkgolide B	OH	OH	H
Ginkgolide C	OH	OH	OH

PHARMACOLOGY

Other ailments and organs which may benefit from ginkgo treatment but which were not discussed in the main body of this booklet are presented below. These physiological effects are due mainly to PAF-acether inhibition.

Liver

High levels of PAF have been detected in the blood of cirrhotic rats. Ginkgolide B shows the ability to counteract some of the negative effects on the body associated with this condition, such as increased cardiac output and decreased arterial pressure.[128]

Gastrointestinal Tract

Bacterial-generated endotoxin may stimulate the production of PAF, leading to ulceration of the GI tract. It is possible that irritable bowel syndrome has a similar pathogenesis and thus would be helped by continued consumption of ginkgo extract.[129]

Kidneys

PAF seems to play a role in the pathogenesis of experimental nephrosis.[130]

Other Physiological Effects

Table 8 summarizes a few more of ginkgo's physiological effects. Most of this research has been done *in vivo*.

TABLE 8
OTHER PHYSIOLOGICAL EFFECTS OF GINKGO

Effect	Description	Possible Health Benefit	References
Anti-oxidant	inhibits free-radicals	prevents damage from free-radicals in tissues (such as the retina) where ginkgo concentrates	131
Reduced tissue oxygen protection	ample oxygen is vital for health of all tissues; stress can restrict blood & oxygen flow	may prevent tissue damage or dysfunction due to lowered oxygen levels in tissues (such as in eyes, ears, and the brain)	132
Cerebral blood flow	opens blood vessels; increases blood flow to some tissues	may increase blood flow to brain; better oxygenation and glucose supply to tissues	133
PAF-acether inhibitor	phospholipid released by human immune cells may play a key role in asthma, brain damage subsequent to stroke	ginkgo extract may inhibit these substances and act as a preventative against asthma; may beneficially affect toxic shock syndrome or septicemia and shock after thermal injury; may reduce brain damage after stroke	134, 135 136 137
Neurotransmitter receptor binding enhancement	possibly counteracts the decreased ability of neurotransmitters to send nerve impulses	may improve memory and awareness where deficient due to aging or functional imbalance	138
Adaptation of excitation of the organ of Corti	helps hearing organ to adapt to noise more quickly	possible protection against hearing loss by noise	139
Retina protective ability	the retina is highly sensitive to free-radical induced peroxidation damage	possible protection of the retina in diabetic patients; loss of vision is a common problem with diabetics	140
Neurotransmitter effects	cerebral muscarinic receptors, norepinephrine turnover increased	reduced muscarinic receptor binding may be contributed to age-related cognitive disorders	141

RESOURCES

Ginkgo Trees Shipped: Forest Farm (503) 846-6963
990 Tetherow Road   Williams, OR 97544

Note: *All of the products discussed in this book are available at local herb stores, health food stores or natural foods markets. The Schwabe product is sold by* **Nature's Way Products** *as* **Ginkgold***.*

To Order Bulk Chinese Herb Products: Nu-Herb 1-800-233-4307
Mayway (415) 788-3646
(ginkgo nuts)

Inspiring Books:
    Louise Hay *You Can Heal Your Life*
    Bernie Siegel Any books; *Love, Medicine and Miracles*
    Paul Bragg Any books, especially *The Miracle of Fasting*
    Svevo Brooks *The Art of Good Living*
    Rene Dubos Any books, especially *Mirage of Health*

Herb Books for Further Study:
    Michael Tierra *The Way of Herbs, Planetary Herbology*
    David Hoffmann *The Holistic Herbal* and others
    Kathi Keville many articles in *Vegetarian Times*
    Richard Mabey *The New-Age Herbal*
    Christopher Hobbs *Herbal Formulas That Work, Vitex, Echinacea, The Immune Herb* and others

GENERAL REFERENCES ON GINKGO

Braquet, P. (ed.). 1988. *Ginkgolides — Chemistry, Biology, Pharmacology and Clinical Perspectives, Vol. 1.* Barcelona: J.R. Prous Science Publishers.

Fünfgeld, E.W. 1988. *Rökan — Ginkgo biloba.* New York: Springer-Verlag.

Agnoli, A., *et al.* 1985. *Effects of Ginkgo biloba Extract on Organic Cerebral Impairment.* London: Eurotext.

CITED REFERENCES

1. Michel, P.F. & D. Hosford. 1988. Ginkgo biloba: from "living fossil" to modern therapeutic agent. In *Ginkgolides*, Vol. 1, ed. P. Braquet, 1-8. Barcelona: J. R. Prous.
2. Bensky, D. and A. Gamble. 1986. *Chinese Herbal Medicine*. Seattle: Eastland Press.
3. Andrews Jr., H.N. 1970. *Ancient Plants and the World They Lived In*. Ithaca: Cornell University Press. pp. 157-167.
4. Shih-Chen, L. 1578. *Pen Ts'ao*. Translated and researched by F.P. Smith and G.A. Stuart, published under the title *Chinese Medicinal Herbs*. San Francisco: Georgetown Press, 1973.
5. Michel & Hosford, *Living fossil*.
6. Shih-Chen, *Pen Ts'ao*.
7. Porterfield, Jr., W.M. 1951. The principal Chinese vegetable foods and food plants of Chinatown markets. *Economic Botany* 5: 3-37.
8a. Perry, L.M. 1980. *Medicinal Plants of East and Southeast Asia*. Cambridge: The MIT Press.
8b. Wada, K., *et al*. 1985. An antivitamin B6, 4'methoxypyridoxine, from the seed of *Ginkgo biloba* L. *Chem. Pharm. Bull*. (Tokyo) 33: 3555-7.
9. Tsuyuki, H., *et al*. 1979. Lipids in ginkgo seeds. *Nihon Daigaku Noj. Gaku. Key. Hok*. 36: 156-62.
10. Porterfield, *Chinese vegetable foods*.
11. Boralle, N., *et al*. 1988. Ginkgo biloba: a review of its chemical composition. In Braquet, *Ginkgolides*.
12. Mitsuki, Y. 1955. Japanese foods. Digestibility and absorption rates of vegetables and fungi. *Fukuoka Acta Med*. 46: 341-7.
13. *ibid*.
14. Perry, *Medicinal Plants*, 160.
15. *ibid*.
16 . Bensky, D. & A. Gamble. 1986. *Chinese Herbal Medicine Materia Medica*. Seattle: Eastland Press.
17. Hooper, D. 1929. On Chinese medicine: drugs of Chinese pharmacies in Malaya. *The Gardens' Bulletin*, Straits Settlements (Singapore) 6 (1): 1-165.
18 Smith, F.P. & G.A. Stuart. 1973. *Chinese Medicinal Herbs*. San Francisco: Georgetown Press.
19. Revolutionary Health Committee of Hunan Province. 1977. *A Barefoot Doctor's Manual*. Seattle: Madrona Publishers.
20. Hsu, H.-Y. 1986. *Oriental Materia Medica*. Long Beach: Oriental Healing Arts Institute.
21. Perry, *Medicinal Plants*, 60.
22. Chang, H.-M. & P.P.-H. But. 1986. *Pharmacology and Applications of Chinese Materia Medica*. Philadelphia: World Scientific.
23. Michel & Hosford, *Living fossil*.
24. Publications and Information Directorate. 1956. *The Wealth of India*, v. IV. New Delhi: Council of Scientific & Industrial Research.
25. Foster, S. 1986. Ginkgo, herb for the future. *Business of Herbs Bull* 4: 3.
26. Dahms, W. 1965. Über Behandlungsergebnisse mit dem Ginkgo biloba-Präparat Tebonin. *Aus Unser. Arb*. 4: 2.

27. HerbalGram: Ginkgo review will be published in 1991, to order: (800) 373-7105.
28. Bauer, U. 1988. Ginkgo biloba extract in the treatment of arteriopathy of the lower limbs. In Fünfgeld, E.W., ed., *Rökan (Ginkgo biloba)*. NY: Springer-Verlag.
29. Locatelli, G.R. & E. Sorbini. 1969. A *Ginkgo biloba* (L.) leaf extract (Tebonin) in the treatment of senile peripheral arteriopathy. *Minerva Cardioangiologica* 17: 1103-8.
30. Weiss, R.F. 1988. *Herbal Medicine*, Beaconsfield, England: Beaconsfield Publishers Ltd.
31. Lagrue, G., *et al.* 1988. Idiopathic cyclic edema; role of capillary hyperpermeability and its correction by *Ginkgo biloba* extract. In Fünfgeld, *Rökan*.
32. Tronnier, H. 1978. Clinico-pharmacological investigations on the effect of an extract of *Ginkgo biloba* L. on the postthrombotic syndrome. *Arzneim.-Forsch.* 18: 551-2.
33. Taillandier, J., *et al.* 1988. *Ginkgo biloba* extract in the treatment of cerebral disorders due to aging; longitudinal, multicenter, double-blind study versus placebo. In Fünfgeld, *Rökan*.
34. Fünfgeld, E.W. & D. Stalleicken. 1988. The clinical effect of *Ginkgo biloba* extract in the case of cerebral insufficiency documented by dynamic-brain-mapping: a computerized EEG evaluation. In Fünfgeld, *Rökan*.
35. Pidoux, B. 1988. Effects of *Ginkgo biloba* extract on cerebral functional activity; results of clinical and experimental studies. In Fünfgeld, *Rökan*.
36. Agnoli, A., *et al.*, pres. 1985. *Effects of Ginkgo Biloba Extract on Organic Cerebral Impairment.* London: Eurotext.
37. Eckmann, F. & H. Schlag. 1982. Controlled double blind study to prove the effectiveness of Tebonin forte in patients with cerebral insufficiency. *Fortschr. Med.* 100: 1474-8.
38. Hemmer, R. & O. Tzavellas. 1977. The cerebral effectiveness of a plant preparation from *Ginkgo biloba*. *Arzneim.-Forsch.* 491-3.
39. Haan, J., *et al.* 1982. *Ginkgo biloba* flavonglycosides—therapeutic possibilities in cerebral insufficiency. *Med. Welt* 33: 1001-5.
40. Hindmarch, I. 1988. Activity of *Ginkgo biloba* extract on short-term memory. In Fünfgeld, *Rökan*.
41. Warburton, D.M. 1988. Clinical psychopharmacology of *Ginkgo biloba* extract. In Fünfgeld, *Rökan*.
42. Dubreuil, C. 1988. Therapeutic trial in acute cochlear deafness. In Fünfgeld, *Rökan*.
43. Meyer, B. 1988. A multicenter randomized double-blind study of *Ginkgo biloba* extract versus placebo in the treatment of tinnitus. In Fünfgeld, *Rökan*.
44. Meyer, B. 1986. A multicenter study of tinnitus. Epidemiology and therapy. *Ann. Otolaryngol. Chir. Cervicofac.* 103: 185-8.
45. Claussen, C.F. 1988. Diagnostic and practical value of craniocorpography in vertiginous syndromes. Fünfgeld, *op. cit.*
46. Haguenauer, J.P., *et al.* Treatment of disturbed equilibrium with *Ginkgo biloba* extract; multicenter double-blind study versus placebo. Fünfgeld, *op. cit.*
47. Lebuisson, D.A, *et al.* 1988. Treatment of senile macular degeneration with Ginkgo biloba extract. A preliminary double-blind study versus placebo. In Fünfgeld, *Rökan*.
48. Drieu, K. 1985a. Multiplicity of effects of *Ginkgo biloba* extract: current status and new trends. In Agnoli, *Effects of Ginkgo*.

49. Michel, P.F. 1985. Chronic cerebral insufficiency and *Ginkgo biloba* extract. In Agnoli, *Effects of Ginkgo*.
50. Leroy, M., *et al.* 1978. Approche clinique et psychometrique en geriatrie. Methodes d'etudes et choix d'une therapeuthique. *Vie Medicale* 28: 2513-19.
51. Eckmann, F. & H. Schlag. 1982. Controlled double-blind study on the proof of the effect of Tebonin forte in patients with cerebrovascular insufficiency. *Fortschr. Med.* 31-32: 1474-8.
52. Arrigo, A. & S. Cattaneo. 1985. Clinical and psychometric evaluation of *Ginkgo biloba* extract in chronic cerebrovascular diseases. In Agnoli, *Effects of Ginkgo*.
53. Weitbrecht, W.V. & W. Jansen. 1985. Double-blind and comparative (*Ginkgo biloba* versus placebo) therapeutic study in geriatric patients with primary degenerative dementia — a preliminary evaluation. In Agnoli, *Effects of Ginkgo*.
54. Cousins, *et al.* (eds.). *Advances, the Journal for Mind-Body Health.*
55a. Michel, *Chronic cerebral insufficiency.*
55b. Arrigo & Cattaneo. *Psychometric evaluation.*
56. Fünfgeld, E.W. & D. Stalleicken. 1988. The clinical effect of Ginkgo biloba extract in the case of cerebral insufficiency documented by dynamic-brain-mapping—a computerized EEG evaluation. In Fünfgeld, *Rökan*.
57. Sitzer, G. 1987. Cerebrovascular insufficiency. II. EEG-long-term studies on the efficacy of Tebonin forte on patients with chronic cerebrovascular insufficiency. *Der Kassenarzt* 17/18, April.
58. Tea, S., *et al.* 1979. Effets clinique, hemodynamique et metabolique de l'extrait de *Ginkgo biloba* en pathologie vasculaire cerebrale. *Gazette Medicale de France* 86: 4149-52.
59. Rapin, J.R. & M. Le Poncin-Lafitte. Cerebral glucose consumption — effect of *Ginkgo biloba* extract. In Fünfgeld, *Rökan*.
60. Schaffler, K. & P. Reeh. 1985. Long-term drug administration effects of *Ginkgo biloba* on the performance of healthy subjects exposed to hypoxia. In Agnoli, *Effects of Ginkgo*.
61. Editorial. 1988. *Ginkgo biloba* extract: over 5 million prescriptions a year. *The Lancet*, December 23/30.
62. Lostre, F. 1988. From the body to the cellular membranes: the different levels of pharmacological action of *ginkgo biloba* extract. In Fünfgeld, *Rökan*.
63. Seyle, H. 1976. *The Stress of Life*, revised edition. New York: McGraw-Hill Book Co.
64. Brooks, S. 1990. *The Art of Good Living*. Boston: Houghton-Mifflin.
65. Ernst, E. 1985. Hemorheological effects of standardized *Ginkgo* extract in vitro. In Agnoli, *Effects of Ginkgo*.
66. Clostre, F. 1988. From the body to the cellular membranes: the different levels of pharmacological action of *Ginkgo biloba* extract In Fünfgeld, *Rökan*.
67. Fünfgeld & Stalleicken, *Clinical effect of Ginkgo.*
68. ADIS Press Limited. 1990. *Ginkgo biloba Extract*. Auckland: ADIS Press Limited, p. 3.
69. *Ibid.*
70. Halliwell, B. & J.M.C. Gutteridge. 1985. *Free Radicals in Biology and Medicine*. Oxford: Clarendon Press.
71. ADIS Press, *Ginkgo.*
72. Drieu, *Multiplicity of effects.*
73. Halliwell & Gutteridge, *Free Radicals.*

74. Pincemail, J. & C. Deby. 1988. The antiradical properties of *Ginkgo biloba* extract. In Fünfgeld, *Rökan.*
75. Chatterjee, S.S. 1985. Effects of *Ginkgo biloba* extract on cerebral metabolic processes. In Agnoli, *Effects of Ginkgo.*
76. Taylor, J.E. 1985. The effects of chronic, oral Ginkgo biloba extract administration on neurotransmitter receptor binding in young and aged Fisher 344 rats. In Agnoli, *Effects of Ginkgo.*
77. Michel, *Chronic cerebral insufficiency.*
78. Hindmarch, I. 1988. Activity of *Ginkgo biloba* on short-term memory. In Fünfgeld, *Rökan.*
79. Michel, *Chronic cerebral insufficiency.*
80. Coier, E. 1988. PAF-acether analogs, platelet activation and BN 52021. In Braquet, *Ginkgolides.*
81. Guinot, P., *et al.* 1988. The clinical effects of BN 52063, a specific PAF-acether antagonist. In Braquet *Ginkgolides.*
82. Bourgain, R.H., *et al.* 1988. Arterial thrombosis induced by PAF-acether (1-0-alkyl-sn-glycero-3-phosphoryl-choline) and its inhibition by ginkgolides. In Braquet *Ginkgolides.*
83. Fink, A., *et al.* 1988. Peripheral blood leukocytes, ischemic heart, PAF, ginkgolide B and leukotriene C4. In Braquet *Ginkgolides.*
84. Tamargo, J., *et al.* 1988. Cardiac electrophysiology of PAF-acether and PAF-acether antagonists. In Braquet *Ginkgolides.* 417-431.
85. Berti, F., *et al.* 1988. BN 52021, a specific PAF-receptor antagonist, prevents antigen-induced mediator release and myocardial changes in guinea-pig perfused hearts. In Braquet *Ginkgolides.* 399-410.
86. Bauer, *Treatment of arteriopathy.*
87. Koeppel, F.W. 1980. Therapy with a vasoactive herbal medicine in hearing defects [in patients] with cervical syndrome. *Therapiewoche* 30: 6443-6.
88. Sprenger, F.H. 1986. Inner ear hearing loss—good results with *Ginkgo biloba*. *Ztl. Praxis* 38: 938-40.
89. Koeppel, F.W. 1980. Tebonin-therapy with old hard-of-hearing people. *Therapiewoche* 30: 6443-46.
90. Meyer, B. 1980. Tinnitus—multicenter study. A multricentric study of the ear. *Ann. Oto-Laryng.* (Paris) 103: 185-8.
91. ADIS Press, *Ginkgo.*
92. Moreau, J.P., *et al.* Absorption, distribution, and excretion of tagged Ginkgo biloba leaf extract in the rat. In Fünfgeld, *Rökan.*
93. Doly, M., *et al.* 1988. Platelet-activating factor, electroretinogram and ginkgolide B—therapeutical prospects. In Braquet *Ginkgolides.*
94. Haydee, E.P., *et al.* 1988. Effect of BN 52021 on the arachidonic acid cascade in the inflamed cornea. In Braquet *Ginkgolides.*
95. Lebuisson, *Treatment of senile macular degeneration.*
96. Csato, M.,*et al.* 1988. Studies on the regulation of murine contact dermatitis by systemic treatment with a PAF antagonist, BN 52021. In Braquet *Ginkgolides.*
97. Tamura, N., *et al.* 1988. Platelet-activating factor, human eosinophils and ginkgolide B (BN 52021). In Braquet *Ginkgolides.*
98. Kian, F., *et al.* 1988. Clinical perspectives of PAF-acether antagonists. Braquet, 1988, *op. cit.*
99. Schilcher, H. 1988. *Ginkgo biloba*: investigation on the quality, activity, effectiveness, and safety. *Zeit. f. Phytother.* 9: 119-127.

100. Drieu, *Multiplicity of effects.*
101. Schilcher, *Ginkgo biloba.*
102. Drieu, *Multiplicity of effects.*
103. *ibid.*
104. *ibid.*
105. Schilcher, *Ginkgo biloba.*
106. Dubreuil, C. 1988. Therapeutic trial in acute cochlear deafness. In Fünfgeld, *Rökan.*
107. Watt, J.M. & M.G. Breyer-Brandwijk. 1962. *The Medicinal and Poisonous Plants of Southern and Eastern Africa.* Edinburgh & London: E. & S. Livingstone Ltd.
108. Personal communication with Schwabe representative.
109. Foster, S. 1990. *Ginkgo.* Published by the American Botanical Council, Austin.
110. Drieu, K., *et al.* 1985b. Animal distribution and preliminary human kinetic studies of the flavonoid fraction of a standardized *Ginkgo biloba* extract (GBE 761). *Stud. Org. Chem.* (Amsterdam) 23: 351-9.
111. Moreau, J.P., *et al.* 1986. Double-blind crossover study for the comparison of the bioavailability of *Ginkgo biloba* extract in tablets and liquid formulations. *Arbeitsbericht der Fa. Dr. Willmar Schwabe,* Karlsruhe.
112. Adis Press, *Ginkgo.*
113. Warburton, *Clinical psychopharmacology.*
114. Felber, J.P. 1988. Effect of *Ginkgo biloba* extract on the endocrine parameters. In Fünfgeld, *Rökan.*
115. Dr. Willmar Schwabe Arzneimittel. [d.m.]. *Tebonin forte* ®(detail manual). Dr. Willmar Schwabe.
116. Schilcher (1988), *op. cit.*
117. Watt &Breyer-Brandwijk, *Medicinal and Poisonous Plants.*
118. Becker, L.E. & G.B. Skipworth. 1975. Ginkgo-tree dermatitis, stomatitis, and procitis. *JAMA* 231: 1162.
119. Mitchell, J.C., *et al.* (1981). Leaves of Ginkgo biloba not allergenic for Toxicodendron-sensitive subjects. *Contact Dermatitis.* 47-8.
120. Radford, E., *et al.* 1974. *Vascular Plant Systematics.* NY: Harper & Row.
121. Bailey, L.H. 1930. *The Standard Cyclopedia of Horticulture.* NY: The Macmillan Company.
122. Rehder, A. 1927. *Manual of cultivated trees and shrubs.* New York: Macmillan Co.
123. Bailey, *The Standard Cyclopedia.*
124. Major, R.T. 1967. The *Ginkgo,* the most ancient living tree. *Science* 157: 1270-73.
125. Drieu, K. 1988. Preparation and definition of *Ginkgo biloba* extract. In Fünfgeld, *Rökan.*
126. Boralle, *Ginkgo biloba: a review.*
127. Wilford, J.N. 1988. Ancient tree yields secrets of potent healing substance. *The New York Times,* March 1, 1988.
128. Fernandez-Gallardo, S., *et al.* 1988. Effect of BN 52021 on the hemodynamics of rats with experimental cirrhosis of the liver. In Braquet*Ginkgolides.*
129. Wallace, J.L. 1988. PAF — and endotoxin-induced gastrointestinal ulceration: inhibitory effects of BN 52021. In Braquet*Ginkgolides.*
130. Egido, J., *et al.* 1988. PAF, adriamycin-induced nephropathy and ginkgolide B. In Braquet*Ginkgolides.*
131. ADIS Press, *Ginkgo.*

132. Schaffler, K. & P.W. Reeh. 1985. Double-blind study of the hypoxia-protective effect. *Arzneim.-Forsch.* 35: 1283-6.
133. Heiss, W.-D. & I. Podreka. 1978. Assessment of Pharmacological effects on cerebral blood flow. *Eur. Neurol.* (Suppl. 1): 135-43.
134. Touvay, C., *et al.* 1985. Inhibition of antigen-induced lung anaphylaxis in the guinea pig by BN 5201 a new specific paf-acether receptor antagonist isolated from *Ginkgo biloba. Agents Actions* 17: 371-2.
135. Braquet, P. 1986. Proofs of involvement of PAF-acether in various immune disorders using BN 52021 (Ginkgolide B): a powerful PAF-acether antagonist isolated from *Ginkgo biloba.* L. U. Zor, *et al.*, eds. *Advances in Prostaglandin, Thromboxane, and Leukotriene Research,* Vol. 16. NY: Raven Press.
136. Etienne, A., *et al.* 1985. In vivo inhibition of plasma protein leakage and *Salmonella enteritidis —* induced mortality in the rat by a specific paf-acether antagonist: BN 52021. *Agents and Actions* 17: 368-70.
137. Anon. 1988 (summer). The ginkgo tree and brain trauma. *Science Focus* 3: 1, 19.
138. Taylor, *Effects of chronic, oral Ginkgo.*
139. Stange, G., *et al..* 1976. Adaptational behaviour of peripheral and central responses in guinea pigs under the influence of various fractions of an extract from *Ginkgo biloba. Arzneim.-Forsch.* 26: 367-74.
140. Doly, M, *et al.* 1985. Effects of oxygenated free radicals on the electrophysiological activity of the isolated retina of the rat. *J. Fr. Opthamol.* 8: 273-7.
141. ADIS Press, *Ginkgo.*